KU-182-938

WITHDRAWN

EFFECTIVE PRODUCT DESIGN AND DEVELOPMENT

HOW TO CUT LEAD TIME AND INCREASE CUSTOMER SATISFACTION

THE BUSINESS ONE IRWIN/APICS LIBRARY OF INTEGRATIVE RESOURCE MANAGEMENT

Customers and Products

Marketing for the Manufacturer   *J. Paul Peter*

Field Service Management: An Integrated Approach to Increasing Customer Satisfaction   *Arthur V. Hill*

Effective Product Design and Development: How to Cut Lead Time and Increase Customer Satisfaction   *Stephen R. Rosenthal*

Logistics

Integrated Production and Inventory Management: Revitalizing the Manufacturing Enterprise   *Thomas E. Vollmann, William L. Berry, and D. Clay Whybark*

Purchasing: Continued Improvement through Integration   *Joseph Carter*

Integrated Distribution Management: Competing on Customer Service, Time and Cost   *Christopher Gopal and Harold Cypress*

Manufacturing Processes

Integrative Facilities Management   *John M. Burnham*

Integrated Process Design and Development   *Dan L. Shunk*

Integrative Manufacturing: Transforming the Organization through People, Process and Technology   *Scott Flaig*

Support Functions

Managing Information: How Information Systems Impact Organizational Strategy   *Gordon B. Davis and Thomas R. Hoffmann*

Managing Human Resources: Integrating People and Business Strategy   *Lloyd Baird*

Managing for Quality: Integrating Quality and Business Strategy   *Howard Gitlow*

World-Class Accounting and Finance   *Carol J. McNair*

EFFECTIVE PRODUCT DESIGN AND DEVELOPMENT

HOW TO CUT LEAD TIME AND INCREASE CUSTOMER SATISFACTION

Stephen R. Rosenthal

BUSINESS ONE IRWIN
Homewood, Illinois 60430

NAPIER UNIVERSITY LIBRARY

AC	ON /40 0763/5
LOC SHMK S 658 · 575 Ros	SUP TCF
CON	PR £33-95

© RICHARD D. IRWIN, INC., 1992

All rights reserved. No part of this publication may be
reproduced, stored in a retrieval system, or transmitted,
in any form or by any means, electronic, mechanical,
photocopying, recording, or otherwise, without the prior
written permission of the publisher.

This publication is designed to provide accurate and
authoritative information in regard to the subject matter
covered. It is sold with the understanding that neither the
author nor the publisher is engaged in rendering legal, accounting,
or other professional service. If legal advice or other expert
assistance is required, the services of a competent
professional person should be sought.

*From a Declaration of Principles jointly adopted by a Committee
of the American Bar Association and a Committee of Publishers.*

Sponsoring editor: Jeffrey A. Krames
Project editor: Karen Smith
Production manager: Ann Cassady
Designer: Larry J. Cope
Artist: David Corona Design
Compositor: Eastern Graphics
Typeface: 11/13 Times Roman
Printer: Book Press

Library of Congress Cataloging-in-Publication Data

Rosenthal, Stephen R.
    Effective product design and development : how to cut lead time
and increase customer satisfaction / Stephen R. Rosenthal.
        p.   cm.—(The Business One Irwin/APICS library of
integrative resource management)
    Includes bibliographical references and index.
    ISBN 1-55623-603-4
    1. New products—United States—Management.   2. Product
management—United States.   3. Design, Industrial—United States.
I. Title.
HF5415.153.R68   1992
658.5'75—dc20   91–36252

Printed in the United States of America
1 2 3 4 5 6 7 8 9 0 BP 9 8 7 6 5 4 3 2

For Linda who waited
and Vanna who watched

FOREWORD

Effective Product Design and Development is one book in a series that addresses the most critical issue facing manufacturing companies today: integration—the identification and solution of problems that cross organizational and company boundaries—and, perhaps more importantly, the continuous search for ways to solve these problems faster and more effectively! The genesis for the series is the commitment to integration made by the American Production and Inventory Control Society (APICS). I attended several brainstorming sessions a few years ago in which the primary topic of discussion was, "What jobs will exist in manufacturing companies in the future—not at the very top of the enterprise and not at the bottom, but in between?" The prognostications included:

- The absolute number of jobs will decrease, as will the layers of management. Manufacturing organizations will adopt flatter organizational forms with less emphasis on hierarchy and less distinction between white collars and blue collars.

- Functional "silos" will become obsolete. The classical functions of marketing, manufacturing, engineering, finance, and personnel will be less important in defining work. More people will take on "project" work focused on continuous improvement of one kind or another.

- Fundamental restructuring, meaning much more than financial restructuring, will become a way of life in manufacturing enterprises. The primary focal points will be a new market-driven emphasis on creating value with customers, as well as greatly increased flexibility, a new business-driven attack on global markets which includes new deployment of information technology, and fundamentally new jobs.

- Work will become much more integrated in its orientation. The payoffs will increasingly be made through connections across organizational and company boundaries. Included are customer and vendor partnerships, with an overall focus on improving the value-added chain.

- New measurements that focus on the new strategic directions will be required. Metrics will be developed, similar to the cost of quality metric, that incorporate the most important dimensions of the environment. Similar metrics and semantics will be developed to support the new uses of information technology.
- New "people management" approaches will be developed. Teamwork will be critical to organizational success. Human resource management will become less of a "staff" function and more closely integrated with the basic work.

Many of these prognostications are already a reality. APICS has made the commitment to *leading* the way in all of these change areas. The decision was both courageous and intelligent. There is no future for a professional society not committed to leading-edge education for its members. Based on the Society's past experience with the Certification in Production and Inventory Management (CPIM) program, the natural thrust of APICS was to develop a new certification program focusing on integration. The result, Certification in Integrated Resource Management (CIRM) is a program composed of 13 building block areas which have been combined into four examination modules, as follows:

Customers and products
    Marketing and sales
    Field service
    Product design and development
Manufacturing processes
    Industrial facilities management
    Process design and development
    Manufacturing (production)
Logistics
    Production and inventory control
    Procurement
    Distribution
Support functions
    Total quality management
    Human resources

Finance and accounting
Information systems

As can be seen from this topical list, one objective in the CIRM program is to develop educational breadth. Managers increasingly *must* know the underlying basics in each area of the business: who are the people who work there, what are day-to-day *and* strategic problems, what is state-of-the-art practice, what are the expected improvement areas, and what is happening with technology? This basic breadth of knowledge is an absolute prerequisite to understanding the potential linkages and joint improvements.

But it is the linkages, relationships, and integration that are even more important. Each examination devotes approximately 40 percent of the questions to the connections *among* the 13 building block areas. In fact, after a candidate has successfully completed the four examination modules, he or she must take a fifth examination (Integrated Enterprise Management), which focuses solely on the interrelationships among all functional areas of an enterprise.

The CIRM program has been the most exciting activity on which I have worked in a professional organization. Increasingly, manufacturing companies face the alternative of either proactive restructuring to deal with today's competitive realities, or just sliding away—giving up market share and industry leadership. Education must play a key role in making the necessary changes. People working in manufacturing companies need to learn many new things and "unlearn" many old ones.

There were very limited educational materials available to support CIRM. There were textbooks in which basic concepts were covered and bits and pieces which dealt with integration, but there simply was no coordinated set of materials available for this program. That has been the job of the CIRM series authors, and it has been my distinct pleasure as series editor to help develop the ideas and facilitate our joint learning. All of us have learned a great deal, and I am delighted with every book in the series.

Thomas E. Vollmann
Series Editor

PREFACE

Until recently, the management of product design and development had been a grossly neglected field of study. Falling somewhere between the established fields of R&D management and production operations management, product development attracted little attention in the management literature, until its widespread competitive significance became apparent. This book is one response to that imbalance. It presents an integrated set of perspectives on the process of new product development, with a joint emphasis on achieving speed to market and customer satisfaction.

The primary orientation of the book is educational: to help the reader appreciate what makes this endeavor exciting, challenging, and, ultimately, successful. However, you should not expect (or even long for) a detailed description of the best way to design and develop a winning product. A more reasonable and constructive expectation is that you will improve your understanding of the issues inherent in this field of management, and that you will become more aware of the strengths and weaknesses of various approaches, given different contexts that might exist or arise.

The book describes typical approaches to product design and development employed by successful firms. Each chapter identifies issues, lessons, and opportunities for improvement. I have attempted to convey a sense for what is generally involved, regardless of the particular kind of product, company, or industry. While a diverse set of examples are used, they emphasize products assembled from discrete parts, rather than products that come from a continuous production process. Company examples concentrate on large-to-medium-sized organizations. When there are significant differences across product and organizational types, I try to identify them.

The seeds for this book came from one of my MBA classrooms at Boston University. In the fall of 1988, I offered a course, co-taught with Peter Lawrence, called Product Design and Manufacturing Policy.

To my knowledge, this course (which had been offered in each of the two preceding years) was the first of its kind to be taught at a major business school. Students enjoyed the course and appreciated the exposure it gave them to managerial aspects of product design, yet many felt that the course was somewhat abstract and that the topics presented were rather disjoint. The students were right. The available written material was sparse and lacked integration. New to the subject myself, I was unable to provide the coherence and integration the subject deserved.

By then, however, I had become intrigued by the strategic importance of product design and the subtlety of issues linking design, development, and manufacturing. Consequently, I decided to explore this field in my upcoming year of sabbatical leave from Boston University. Coming from the field of operations management, I was especially interested in describing and analyzing the gaps in execution that seemed to be so common as companies attempted to turn ideas for new products into tangible market offerings. This book is the product of that interest.

Looking back, I now see this book as a new product introduction effort. It began with an idea for an extensive research project, presented to the steering committee of the Boston University Manufacturing Roundtable, a research center sponsored by more than a dozen large manufacturing companies in different industries. Robert L. Badelt, chairman of the committee (and assistant vice president, Operations, Northern Telecom, Inc.) challenged me to complete the proposed research project in 18 months rather than the proposed 3 years. Responding to this pressure for greater speed to market, I proposed restructuring the project to include working members from the companies that would be studied instead of the more traditional use of only faculty and graduate students. That idea proved to be a critical design decision.

Eighteen months later, our research project was completed and we had assembled the material that would form much of the core of this book. Our project team had engaged in a form of action research in which companies collaborated with a university in the pursuit of timely knowledge on a topic of joint interest. Our cross-industry research had explored product design and development from the first formal efforts at idea justification and conceptual design; through detailed design, development, testing; to the beginning of customer delivery from a volume production facility. We studied how companies had approached this process, what they experienced, and how they subsequently tried to improve what they were doing. Ultimately, we wanted to capture managerial insights that would continue to be relevant for a wide range of manufacturers.

We pursued this subject in depth in seven companies: AGFA Compugraphic (now part of Miles Incorporated, Agfa Division), Amdahl Corporation, GE Aircraft Engines, The Interlake Corporation, Motorola Paging Division, NeXT Inc., and Northern Telecom. For each company, we selected a single completed new product introduction effort and prepared, under faculty direction, a written case history, based for the most part on research performed by members of the business unit under study. The emerging case studies, enriched by a series of five two-day workshops attended by the entire project team, formed the basis for a series of working papers that presented comparative findings on common themes of managing new product introduction.

During that 18-month period, I had the pleasure of working with a number of outstanding individuals. Merrill Ebner, Professor of Manufacturing Engineering at Boston University, was my academic partner in planning and designing the research, recruiting the team, reviewing interim research products, and (subsequently) presenting findings at meetings of the Roundtable and at a conference we developed for the business community. The following company executives were designated research fellows of the Manufacturing Roundtable and served as dedicated part-time members of our project team (each contributing from 30 to 50 days during the course of this project):

Douglas Boike—Vice President, Temple, Barker and Sloane

Jerry Dehner—Director, Strategic Manufacturing Development, Northern Telecom, Inc.

Donald Gregory—Manager, CIM Technology, GE Aircraft Engines

Frank Lloyd—Vice President and Director, Subscriber Products and Systems, Motorola, Inc.

K. C. Venugopal—Director, Advanced Manufacturing Engineering, Amdahl Corporation

Tony Whitton—Manager of Standard Product Sales, The Interlake Corporation

These people started off as customers when we recruited them for our research effort, became associates during the project, and remain as friends. Liza Gentile, now with NeXT Inc., also did an outstanding job in conducting an early case study on NeXT, under my direction, while she was a student at Boston University. The references in this book to the written contributions of these people is but a small part of the value that each added to our joint effort and to my ongoing learning. None of this would have been possible without the openness and support of the companies we studied. They provided us with rare inside views of a core business process and, by allowing our industry collaborators to work on this project, provided resources that money could not buy.

This research project generated a set of perspectives on effective product design and development, which I then refined and extended at other opportunities that arose during the past two years. I would like to thank John Ettlie for inviting me to contribute a chapter to his recent book, Gerald Susman and Joseph Blackburn for the conferences they hosted, the speakers and participants at the "New Product Speed to Market" conference we held at Boston University, and the companies with whom I have consulted. These occasions brought countless interesting discussions with others and gave me an opportunity to develop my thoughts.

The Boston University Manufacturing Roundtable provided resources, access, and an important sounding board in support of our research project. I would like to thank Bob Badelt, Stu Christie, Dick Renwick, and Bill Smith—corporate members of the Manufacturing Roundtable steering committee—for their special personal interest and involvement with our research and conference. Bob Badelt also provided helpful suggestions on several chapters of the book manuscript. I am also grateful to my colleagues Dick D'Entremont and Fred Scott, who—as Executive Directors of the Roundtable during the past several years—encouraged me to pursue this work. I greatly appreciated a grant from the Alfred P. Sloan Foundation that covered some of the expenses associated with the preparation of this manuscript, and use of the facilities and support services of the Amos Tuck School of Business Administration at Dartmouth College (where I enjoyed spending part of my sabbatical year).

All of my colleagues in the Operations Management Department at Boston University provided daily doses of encouragement during the research project and the writing that followed. Specific thanks are due to Tom Vollmann, who helped me envision the form the book would take and offered constructive comments on the entire manuscript, and to Jeff

Miller, who, for the past decade, has sparked my interests in manufacturing, technology management, and product design.

Mohan V. Tatikonda, a doctoral student at Boston University, was a true colleague during the past two-and-a-half years. He participated in the original research design, was an active and valued member of our research team, wrote one case study, and collaborated with me on two working papers. His comments on the drafts of several of the chapters of this book were, as usual, incisive.

From the first session of organizing our research on this subject to his reading of the entire final manuscript of the book, Jerry Dehner has been a valued colleague and friend. With dedication and good cheer, Lisa J. Morin helped me meet my publication deadline and produce a readable manuscript by reviewing and updating all chapter drafts. Lisa and I learned to play quite a duet on the word processor! In reviewing drafts of some of the chapters, Artemis March and Janet Bond Wood taught me much about saying what I mean—and meaning what I say.

As the material in this book is aimed at promoting learning in both classrooms and company offices, I wish to thank my students at the Boston University School of Management and participants in executive education programs and conferences in which I have presented portions of this material. I have benefited from their comments and have been pressed onward by their questions.

Stephen R. Rosenthal

CONTENTS

CHAPTER 1

DESIGNING AND DEVELOPING PRODUCTS MORE EFFECTIVELY

New products determine the future of manufacturing companies. Without well-designed, effectively developed new products, a company's prosperity is limited. A lack of successful new products can even threaten companies whose prior product success was legendary, two recent examples being Wang Laboratories' information-based office systems and Polaroid's instant photography. Counting on the continued rewards from existing products can be a recipe for ruin. Although a stream of incremental improvements to existing products may forestall disaster, it is usually not long before competitor initiatives and shifting customer needs make a firm's existing products obsolete.

Companies try to respond to market changes and competitive pressure by introducing new products, but doing this with success requires more thought and effort than many have invested. Most initial ideas for new products fail, some by being canceled during the process of design and development and others by being introduced and then rejected in the marketplace. To be successful, a new product must offer customers distinctive value through some combination of its utility, style, price, and availability, in comparison with other existing products.

Throughout history, people with vision and skills have recognized the demand for new and better products and have looked to the untapped potential of technological possibilities. Occasionally, the resulting products have been so unique and timely that they became the precursors of entire industries. The Model-T Ford and the first product of the Haloid Corporation (later to become Xerox Corporation) were two such radical breakthroughs. In more recent years, the introduction of Sony's Walkman and Sun Microsystem's workstation applied existing technology to new markets and set the stage for other companies to follow.

More typically, however, new products displace existing products and shift the basis of competition in established industries and markets. The use of new materials and advanced manufacturing processes has led

1

to the design and development of successful products of all types: auto-mobiles, home appliances, toys, and footwear, to name a few familiar examples. Such new products may propel a start-up firm to successful patterns of growth, or allow an established firm to hold or enhance its market share. For more mature companies, the process of product design and development can become a vital renewal activity, often leading to the modernization of outmoded marketing, design, manufacturing, and dis-tribution processes.

The process by which a product idea becomes a commercial reality thus brings regular life-and-death challenges for almost any manufactur-ing company. Companies are forced to respond to these challenges by changing their way of product design and development to be more ef-fective. The term *effective* means doing the right thing—in this case, designing a product that, once manufactured and delivered, will exceed customer expectations while meeting company objectives with respect to revenues, profits, or market share, and setting a base for the future.

Unfortunately, manufacturing in the United States has had a pattern of unsatisfactory outcomes in trying to meet these traditional objectives. The widely disseminated report *Made in America*, based on extensive analysis of U.S competitiveness in several industrial sectors, pointed to "a problem in productive performance . . . manifested by sluggish pro-ductivity growth and by shortcomings in the quality and innovativeness of the nation's products" (Dertouzos, Lester, and Solow, 1989, p. 166). Companies that cannot overcome these problems of performance will eventually fail. Industries where this is an endemic problem will continue to falter.

While certain aspects of new product introduction (NPI) are receiv-ing increased attention, the progress in many companies has been slower than expected. Currently, there is no shortage of strategic ideas to apply. Important and useful concepts—such as "competing against time" (Stalk and Hout, 1990), "the attacker's advantage in innovation" (Foster, 1986), "dynamic manufacturing" (Hayes, Wheelwright, and Clark, 1988), and "leadership for quality" (Juran, 1989)—have been articulated in recent years. Increased familiarity with such concepts, coupled with the recent major restructuring of many manufacturing companies, has promoted new orientations to product design and development. Despite this prolif-eration of strategic initiatives, many companies have found the achieve-ment of meaningful levels of improvement to be difficult. For example, the 1990 Manufacturing Futures Survey found that large U.S. manufacturing business units have identified faster product development as a top priority, yet this is an area in which their competitive advantage was comparatively weak (Miller and Kim, 1990).

COMPLEXITY AND CENTRALITY

The first step in making substantial improvement in this area is to acknowledge that product design and development is a complex business process with central ties to contemporary management concerns. Three such aspects of its complexity and centrality can be seen in connections with overall business strategy, time-based competition, and cost structures.

How Product Design Supports Business Strategy

Product design and development is an ongoing company activity and ought to be consistent with the overall business strategy of a company. Various business strategies call for a different emphasis in the design and development of products, and a company's business strategy ought to be compatible with its distinctive competence in the introduction of new products.

If, for example, a company chooses to compete based on providing extraordinary value and responsiveness to its customers, it will have to deal simultaneously with issues of cost, quality, and delivery time in its new product deliberations. Detailed decisions about the form, fit, and function of a new product and the characteristics of its manufacture must include all of these considerations. The cost considerations alone are complex, as are those of quality and delivery time. Functional capabilities, reliability, and ease-of-use are only several of the many dimensions in which a customer might assess quality. Delivery time depends on the supply of materials and components and on the reliability and flexibility of the manufacturing system, both of which are determined largely through decisions made during a product's design.

Consider the implications of the business strategy of pursuing multiple market segments. This strategy aims to provide a comfortable profit margin from a variety of products that are highly differentiated to meet the specialized needs of different customers. A close and accurate orientation to the needs of these customers and the ability to design products to control the costs of variety are essential in this endeavor. A company with this strategy must be skilled in designing products either modularly or in families, with special attention to keeping the number of parts and suppliers as low as possible.

Finally, consider the business strategy of continuously improving a company's existing products. This strategy requires a highly productive process of product design and development that is well-understood by all involved. The improvement of existing products and the introduction of new products must become a seamless ongoing activity. Product design must be conducted with a planned rhythm (and, usually, with short product development cycle times). The design approach needs to emphasize product extensions and product families. In many industries this requires the ability to combine patterns of ongoing, incremental improvements in technology with occasional, carefully selected, and well-timed technology "jumps." Flexible manufacturing capabilities can be essential to accommodate these rapid and frequent product changes.

Illustrative business strategies are supported by a common set of capabilities:

- Offering new products that anticipate changing customer requirements, and doing so in a timely fashion.
- Solving cross-functional problems using multiple design criteria.
- Effectively cataloging and introducing new technology.
- Routinely capturing and reusing product/process design data.
- Following well-defined routine processes of design and development.
- Continuously improving organizational structures and processes.

Such capabilities require an integrative approach, including compatible organizational structures and practices, and appropriate human resources, managerial styles, and technologies. Operational complexity of this sort is inherent in designing and developing products, regardless of the industry or the size of the company.

How Product Design Requires Speed and Discipline

Much has been written in the last several years about time-based competition. In manufacturing, the initial focus was on the process of taking orders, manufacturing individual units, and delivering products. More recently, notions of time-based competition have been expanded to encompass product design and development. Excelling in both aspects of time-based competition calls for similar capabilities including: streamlining the operations to eliminate activities that do not add value; doing the right thing right the first time; building the capacity needed to handle surges in demand; and working more closely with suppliers and customers (Blackburn, 1991). "Speed to market" has become a watchword in many industries, as more and more large U.S. manufacturers institute corporate, companywide initiatives in time management. Whenever we hear of a company that slashed a huge percent off the time to commercialize technology in a new product, the message that "faster is better" gets reinforced.

Time management for new product introduction now runs the risk of oversimplification. As managers are pushed to achieve sizable improvements in product development, they have become absorbed with the element of time, but speed is only one of the measures of performance that must be managed. Elements of quality and cost will always be important too. In product design and development, achieving speed with purpose, and avoiding reckless speed, requires careful management of the whole process.

The realization that we operate in a world of multiple objectives should be familiar, although not comfortable, to most managers in manufacturing companies. With respect to the concept of quality, management around the globe has come to understand that cost and quality are not opposing objectives. The movement toward total quality management (TQM), furthermore, embraced the notion that everyone's job is to achieve increasingly higher levels of quality. Leading manufacturers have moved beyond quality as a sole objective and now include the reduction of time to market. Factoring in this new "time" variable without losing sight of the rest of the multiobjective equation is one of the major challenges facing those working in today's manufacturing environment. Although all of this cannot simply be reduced to obvious formulas for success, disciplined, streamlined processes of design and development can result in shorter lead times and yield products with higher quality and lower cost.

How Product Design Affects Costs

Companies that are restructuring to reduce their overall costs will eventually look to their product design and development activity, since this can be a source of considerable cost. As outlined in Figure 1–1, product

---

FIGURE 1–1
Product Design and Development Affects Costs

	Cost Elements

Inputs to
   product design
   and development
	{ Project number, duration, and structure
	Use of human resources
	Use of design technologies

Generated by
   product design
   and development
	Materials
	Parts
	Labor
	Equipment
	Facilities
	Methods
	Yields
	Repairs

---

design and development activities affect a company's cost structure through the cost of the project itself, and through the resulting product and manufacturing process. The product development cost can be reduced by limiting the number and duration of such projects, simplifying their structure, and being more productive in their execution. Unit costs of new products depend significantly on decisions made during product design. In particular, unit costs can be reduced by efficient manufacturing, which can be made a priority in the early conceptual stage of product design and development.

Getting manufacturing, engineering, and other costs under control is a competitive requirement for many companies. In pursuing this objective of cost reduction, managers need to remain oriented to the effectiveness of product design and development, not only to its costs. When managers choose not to pursue ideas for a new product, they should do so in an informed manner—based on the belief that the ideas, once developed into an actual product, will not justify the costs and risks that need to be incurred. Similarly, product ideas that do appear fruitful should be viewed as investments. In allocating resources to develop such potentially profitable product ideas, they should consider the lost revenues from introducing a product that fails to satisfy customer requirements or is late and misses a narrow market window. Not to consider such factors can turn out to be very expensive ("penny wise and pound foolish").

THE VALUE OF OPERATIONAL PERSPECTIVES

Managers who seek to improve this complex business process should adopt a comprehensive point of view, in which operational realities are emphasized. Only then can they ensure that their interventions are substantive and effective. They should also instill among the participants in the introduction of new products, shared fundamental perspectives on this subject. Toward this end, the NPI process can be seen as a series of *projects*, each of which is made up of a series of *decisions* that combine to *transform* an initial product into a marketable physical reality. Thus, product design and development is both a collection of discrete projects and an ongoing business activity. Each of these perspectives is valid and provides different insights.

This book (and the research upon which it is based) concentrates on the single NPI project for two reasons: First, analysis at the project level is both potent and practical. Potency comes from achieving significant benefits by changing how product design and development activities are conducted. This is practical at the project level because we now know enough about cross-functional management to address specific issues and opportunities. Second, managing NPI projects well is a basic building block for successful product innovation over time. The development of strategy and synergies across projects is extremely important, but it is no substitute for effective project planning and execution. Instead, the long-term success of families of products can only be achieved if the capability to execute individual projects is outstanding. Furthermore, much of the organizational learning and associated efforts at continuous improvement, so important for long-term success, takes place at this critical level of the individual product development project.

A company's long-term success is usually based on the pace, direction, and acceptability of its stream of products over time, rather than on any of its single products. Viewed strategically, then, product design and development is an ongoing process, a continuing series of discrete projects with important continuing relationships to other business activities such as R&D, marketing, manufacturing, and field service. Although this book focuses on the planning and execution of individual NPI projects, it also explores matters of strategy and describes ongoing relationships with other business activities.

By understanding the kinds of *decisions* that are associated with product design and development, one gets a sense of the operational re-

FIGURE 1–2
Decisions Affecting Product Design and Development

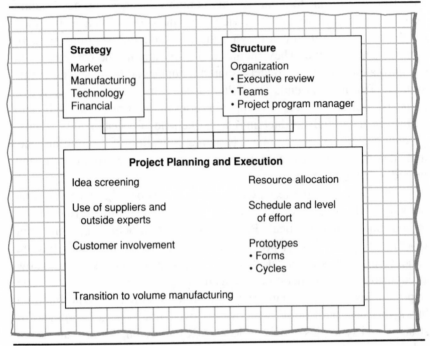

quirements of this activity, no matter how the individual decisions are structured or who participates in making them. Figure 1–2 lists some of these areas of decision making. Senior management sets the stage with strategic decisions about the markets and technologies to pursue, the relative emphasis to place on internal manufacturing capabilities, and the structure of the NPI process. Senior management also participates selectively in the review of proposals for new products, then on the progress with respect to any particular new product initiative. Product or functional managers (depending on the organizational structure) decide on the critical success factors for a new product development effort and project managers try to resolve trade-off issues as they arise in the course of the design effort. Experts who work on these design and development projects are responsible for detailed decisions about the form, fit, and function of the product and how it will be manufactured, delivered, and

FIGURE 1–3
Forms of a Product Design

Form	Description
Concept paper	Preliminary qualitative description of intended product
Sketch	Rough drawing of a product or component
Blueprint	Precise drawing of a product or component
Physical model	3-dimensional representation of shape and exterior appearance of product
Simulation model	Programmed representation of product layout or functions
CAD file	Electronic representation of parts/product geometry
Design release bulletin	Full description of product for use in designing the manufacturing process
Bill of materials (BOM)	Precise list of all parts/components of end product
Process plan	Detailed description of how product is to be manufactured
Service plan	Description of field service requirements (such as replacement parts, service delivery standards, technical support procedures, test equipment)

serviced. Although no one person is responsible for the consistency of these decisions, it is usually vitally important that this happens.

Another way to appreciate the operational nature of the NPI process is simply to ask: What is a product design? The answer will probably depend upon who is being asked and it will relate to the sequence of forms taken by the core design and by its ancillary elements. As shown in Figure 1–3, both the form and the information content of a *product design* evolve in conjunction with the design and development decisions. Understanding how a product design achieves these forms, the issues at stake in considering design trade-offs in terms of these forms, and who ought to participate are some key prerequisites for improving the effectiveness of the overall process.

SCOPE AND CONTENTS OF THE BOOK

This book aims to promote more effective management of product design and development by addressing the inherent complexity and operational reality of the subject. It deals with management processes and the nature of design and development in sizable companies, and it addresses the dual requirements of cutting development lead times and achieving customer satisfaction. Building on the preliminary perspectives introduced above, it offers concepts, findings, and illustrations to guide management action.

Here we refer to "new" products as ones that are different enough from existing products to require formal screening and selection of the underlying concept, and deliberate planning and disciplined execution. Such situations include major extensions to an existing project, as well as the introduction of ones that are more unique. Accordingly, readers interested in major product extensions should not be put off by our frequent reference to "new" products (or NPI). Semantics aside, we are dealing with the managerial realities of substantial product design and development in manufacturing companies. This book does not focus on the pursuit of minor product enhancements—which are handled through an ongoing process of engineering change orders (ECOs). We do not deal with these minor enhancements here only because they are accomplished through a separate, less elaborate, and more routine process than our core subject. Ongoing customer satisfaction relies on the effective management of this enhancement process. Experience that is thus gained needs to be routinely assessed in the design and development of new products.

We examine product design and development from several viewpoints, each of which is important to the overall effectiveness of the process. These points of view are presented graphically in Figure 1–4.

Any analytic framework involves choice and this is no exception. Several important ingredients of NPI success are notably missing from Figure 1–4: an orientation to the customer, the involvement of suppliers, and the accessing of important technical skills and managerial experience. This is neither an oversight nor an omission, but simply the result of a decision to use other building blocks for our analytic framework. These other important integrating topics are discussed throughout this book, rather than being featured within any single chapter heading.

The book has three parts:

Part 1.   Conceptual foundations for understanding the challenges and opportunities.

Part 2.   Case illustrations that cut across the concepts.

Part 3.   The broader context of NPI: institutional enablers, organizational linkages, and continuous improvement.

FIGURE 1–4
Integrative Perspectives on Product Design and Development

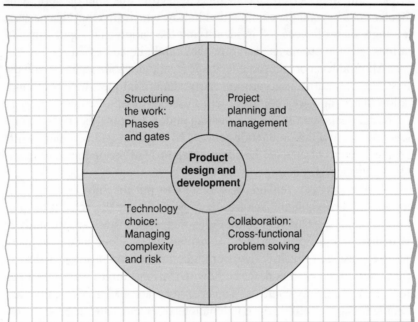

Part 1 (Chapters 2–6) presents concepts and frameworks that facilitate a comprehensive understanding of the managerial dimensions of NPI, covering each of the perspectives shown in Figure 1–4. This material has been derived from recent research by the author and others and draws upon timely examples and relevant literature.

Chapter 2 looks at a basic dilemma in structuring the work of product design and development: management can make this activity more routine by being more explicit about how it is to take place, but the work itself must remain dynamic and flexible for it deals with risk, uncertainty, and rapid changes in the organization's external environment. Chapter 3, looking more closely within a single project, describes the nature of time-based management in the context of product design and development. It identifies the range of critical success factors for such a project and addresses issues of setting and achieving targets.

Product design and development is accomplished through team problem solving. Chapter 4 explores this subject and identifies challenges in managing high-performance teams engaged in this kind of work. This chapter goes beyond the common but hollow conclusion that "teams are a key ingredient of success," in identifying the variety and significance of various "people problems" and how they might be addressed.

All products contain some technological innovation as does even the most basic manufacturing process. Any venture into technological innovation that is new to a company will raise questions of choice and risk. Chapter 5 deals with issues in selecting and developing technology for a new product and the production process by which it will be made, while Chapter 6 concentrates on the adoption and use of technologies to facilitate the product design process itself.

Two tasks then remain. The first is to put the subject of product design and development back together again, having dissected it for purposes of analysis. Our vehicle for doing that is the historical case study of particular NPI projects. In Part 2 (Chapters 7–10), four such case histories are presented, each drawn from a company in a different industry. These narrative chapters describe how the various topics were integrated in different company contexts. The case histories also provide a practical vehicle for appreciating how combinations of external and internal factors can impact the NPI process and challenge its management. The cases were selected to illustrate different themes of broad interest and significance, namely: a start-up company; a large company that had experienced problems with a former product offering and had to make dramatic changes to their NPI process; a company pushing the frontiers of product quality and experimenting with global design of a new product; and a company developing a very complex product in a regulated environment. Each of these case history chapters is concluded with a brief commentary that highlights the relevance of that case to the concepts presented in Part 1.

Part 3 (Chapters 11–14) takes on the remaining task of exploring the broader context within which products are designed and developed. These contextual factors—which have been mentioned earlier in Parts 1 and 2 but require more extended discussion—include a number of enabling capabilities and several ongoing business functions that support product design and development in fundamental ways. Chapter 11 covers the key enablers: total quality initiatives, human resource development, and information technology. Chapter 12 addresses linkages between product design and development and the ongoing activities of manufacturing process development, marketing and field service; NPI activity must be properly integrated with these other processes, even though the day-to-day connections may appear to be rather remote. Chapter 13 looks to the future and discusses emerging issues and an agenda for the continuous improvement of the process of new product introduction. Chapter 14 presents concluding observations on how to cut lead time and increase customer satisfaction.

Product design and development has never been more important as an area for attention by managers and technical experts alike. This fascinating and complex business terrain is normally obscured by its many crossings over traditional functional boundaries. This book offers a practical, comprehensive, and operational view of that subject—an integrative approach that aims to encourage effective management of manufacturing companies in the years ahead.

PART 1

UNDERSTANDING THE CHALLENGES AND OPPORTUNITIES

Effective product design and development begins with a full understanding of the associated activities and how they are conducted, integrated, and managed. The chapters in Part 1 provide conceptual foundations for understanding a wide range of managerial challenges and opportunities. While the central subject of these five chapters remains the same, the perspective shifts from formal structures of review to project management, team development, technology choice and risk, and finally to the activity of design itself. Taken as a whole, Part 1 provides an integrated managerial view of product design and development.

CHAPTER 2

STRUCTURING THE WORK: PHASES, GATES, AND SIMULTANEOUS ENGINEERING

Product design and development projects are part of the innovation process and need to be structured to provide appropriate opportunity for management review. To increase the odds of success through innovation, management must strive for consistency and discipline in the way that design and development activities are conducted. This chapter describes how most large companies, and many small ones, structure the work of NPI (new product introduction) projects. Major improvements to *existing* products, as distinguished from the introduction of essentially *new* products, will require similar managerial mechanisms for shaping the work, reviewing progress, and resolving issues. For simplicity of expression, however, we will continue to refer to the NPI process throughout this chapter.

Product design and development projects are inherently risky, strategically important, and heavy consumers of scarce resources. Senior management needs to decide, for example, how much resource support to provide for any such project, when to terminate a project that is in trouble, and how to cope with unanticipated events and unresolved problems. Projects that sound exciting when initially proposed need to be viewed from time to time to see whether they are meeting their technical and business objectives.

A project's overall success depends on achieving significant added value for the customer. This, in turn, requires some acceptable combination of technological success (making a product with appropriate technology that works as intended) and market success (achieving adequate profits and/or market share). Most attempts to develop a new product will experience some difficulty in meeting expectations in one or both of these dimensions. Projects must be structured to promote the discipline that brings such difficulties to the surface as early as possible.

Senior managers in larger organizations can become involved in product design and development in a constructive way through a well-conceived formal structure of review. Executives in small organizations naturally spend considerable time on issues of product design and development and cannot exploit the simplicity of their organizational structure. (See the Chapter 7 discussion of the NeXT Corp. for an extended example of this situation.) Regardless of the industry, as a company grows, and there are more competing demands on everyone's time, management tends to add formal structure to shape the work of NPI and its review.

This chapter addresses *three* questions:

- What are the common formal approaches to structuring NPI projects?
- What is the practice of simultaneous engineering and why is it important?
- How should management think about these questions of overall NPI structure?

But why address questions of overall project structure and formal management reviews at the beginning of this book, rather than somewhere in the middle or at the end? The answer is simply this: an understanding of the way that the NPI process tends to be structured is useful background for addressing issues of project management, team functioning, and technological choice raised in subsequent chapters.

DESCRIPTION OF THE NPI PROCESS

To begin, we must put the NPI process into the broader context of research and technological innovation. As shown in Figure 2–1, the earliest steps in an innovation sequence are inventions or findings from exploratory scientific research, followed by more applied forms of research aimed at aspects of technological progress deemed to be worth pursuing. Such applied research is usually justified in terms of a perceived general need or market opportunity. These "upstream" innovation activities have their own designated resources and management challenges quite independent of any related and subsequent product design and development activities.

Our focus on product design and development, then, begins after a specific target market has been identified and a product idea has been

FIGURE 2–1
Research versus Product Development

	Type of Activity	Purpose	Primary Measure of Success
Upstream	Exploratory research ("basic")	Creating long-term technical capabilities (invention)	Degree of invention, novelty
	Short-term research ("applied")	Improving current products/ processes	Technological progress
Downstream	Development (product design, testing, and initial production)	Commercialization	Satisfying customer requirements

screened and accepted. We do not discuss the "upstream," research-based stages of innovation because they are usually managed separately from the introduction of new products. Our emphasis, instead, is on the deliberate and targeted "downstream" program of innovation aimed at the commercial offering of a particular new product. The initial point of interest, therefore, is the formal recognition by senior management of a need or an opportunity that might be satisfied by the design and development of a new product. This point is easily identified in practice because it is then that an NPI project is launched, leading to activities directly associated with designing, developing, testing, and manufacturing a new product.

Formal processes for managing new product introductions exist in most large manufacturing companies, and general summarized versions have begun to appear in the management literature (see, for example, Cooper, 1990). While companies may have different names for the phases of their NPI process, they usually map nicely into five phases:

- Idea validation.
- Conceptual design.
- Specification and design.
- Prototype production and testing.
- Manufacturing ramp-up.

Smaller companies with products that are not highly complex, also seeing the need for more structured approaches to product design and development, are following the notions described below. For example, Interlake Conveyors, a manufacturer of industrial shelving and material handling products, moved responsibility for product design and development from the corporate engineering group to their plant and developed systems and procedures to improve the management of that process (Whitton and Cook, 1990).

Without getting into questions of how these phases are performed and the relationships among them, we can describe the content of the work to be done. We do this in terms of the traditional operations management model: inputs being converted to outputs. This framework is chosen because it helps us to appreciate the inherent issues of management review and control at each phase. The framework is summarized in Figure 2–2. The definitions and terminology used in this chapter are an agglomeration from several different companies and are intended to be illustrative of current practice among companies and industries.

FIGURE 2–2
Overview of NPI Phases and Gates

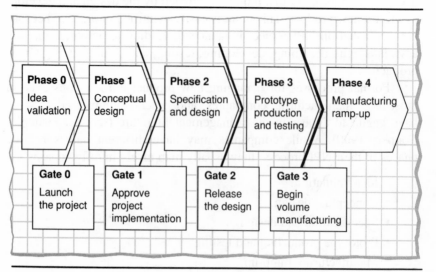

Idea Validation (Phase 0)

The idea validation phase precedes the formation of a formal NPI project. We include it here for completeness, but, since it precedes the management of product design and development, it is not of major interest to us elsewhere in this book. The idea validation phase encompasses the identification, screening, and initial refinement of an idea for a new product. Depending on the industry and a company's innovation strategy, this initial stage of a new product introduction may be triggered by an initiative of senior management, a formal planning exercise by an ongoing group charged with this responsibility, or simply an idea championed by a single individual within the company—or a customer. In a small company, senior management might naturally become aware of opportunities for new products, but in larger companies, it might have to "trickle up" through the organization.

Senior management will often informally launch the idea validation phase of a new product by assembling a small planning team with representation from marketing and technical fields. Their charge, generally speaking, is to try to match some particular market opportunity with available technology. The market opportunity emerges first when a new product idea is generated and explored as a product possibility. Responsibility for the creation of new product ideas may be vested with some particular group, or in companies where innovation is strongly emphasized; there may be encouragement for anyone with such an idea to raise it for consideration. Then, if a product idea successfully passes an initial screening review, in comparison with other such possibilities, it is considered further. The term *idea champion* is sometimes used for the person who attempts to make a persuasive case in any such instance.

In some high-technology companies, an engineering group is designated to scan the market to identify the need for increased performance beyond the scope of other planned products. The time horizon for such market scanning needs to be long enough to encompass a full product development cycle. Clearly, it would be too late to wait for the sales of an existing product to decline as a signal to start to think about the development of a new product. In the large-scale mainframe computer market, the engineering group charged with this kind of scanning may need to look five to seven years ahead.

3M is a good example of a corporation that encourages its employees to generate ideas for new products. In such a situation the original idea champion is provided a limited amount of resources (and a small fraction of their normal working hours) to pursue the idea far enough to make a formal proposal that the idea receive further investigation and support. This is more likely to be the case when a company already has a very large number of products and a policy that a significant fraction of sales should come from products that are only a few years old.

In any event, the case for a new product is structured in terms of identifying the two-dimensional space—market and technical performance—into which the new product idea fits. Typical considerations at this point would include:

1. A definition of the target customer and an overview of the customers' unmet needs.
2. A sense of the technological options that exist to develop a product that meets those needs.
3. The issues regarding suitable manufacturing technologies and facilities.
4. The general market potential for a product of this type.
5. The competitive advantage of the company in pursuing the introduction of such a product.
6. The resources, both dollars and personnel, that would be needed to proceed to develop the product concept.

Draft concepts and plans for approved new product ideas are then prepared in the form of a business opportunity plan (BOP), sometimes with the help of a designated product manager (usually drawn from a marketing or engineering design orientation). The team that prepares the BOP may be small but typically includes expertise from the fields of marketing, product design, and manufacturing. The contents of a BOP will differ somewhat from company to company but would normally include:

- Forecasts of technology trends and customer preferences (including relevant market research data).
- An analysis of the competitive status of the company if this product is successfully introduced.
- An outline of major design requirements for the product (such as features, functions, reliability, size, weight) and the associated manufacturing process (that is, quality/yield, delivery time).

- An analysis of competitive threats and responses.
- A marketing plan including sales and market share forecasts.
- A product development plan (costs, timing, and organizations to be involved).
- A financial analysis to justify the new product introduction.

The business opportunity plan represents the overall goal of the company and should contain internally consistent objectives across the participating functional areas which would include marketing, design, manufacturing; support organizations such as finance, legal, and quality; and others that a company would normally specify. In some instances, the BOP would also recommend a project manager (sometimes called the program manager) who would be in charge of this NPI effort, should it be approved.

Conceptual Design (Phase 1)

At this next phase the concept expressed in the idea validation phase is elaborated and extended with an eye toward demonstrating the business feasibility of the new product. (Some companies call this the *feasibility phase*). Ideas are made more concrete as commercial specifications are identified for the new product, including the performance and aesthetic attributes of the product and its price. Such commercial specifications set customer expectations and guide the work of product designers and manufacturing engineers who then proceed to identify attributes of product's form, fit, and function. Technical feasibility is explored by defining the product architecture (individual elements and their interrelationships), while identifying broad design options and trade-offs, possible technical "show stoppers," and the associated risk. To the extent that this product will require that a new manufacturing process be developed, such needs will be described.

Marketing feasibility is assessed in terms of business strategy, marketing and sales objectives, marketing tactics, and the associated resource requirements and product launch schedule. Customer acceptance and sales estimates are also assessed at this phase through various types of market research, concept testing, and focus groups often through the use of product mock-ups and models. The normal role of marketing in this initial phase of new product introduction is discussed further in Chapter 12. However, when a product is being designed to create a new market, traditional market analysis can be less valuable than the visionary zeal of the product planners. Achieving a balance between traditional market research and visionary zeal is a key factor for long-term success.

The output from the conceptual design phase is usually a document that demonstrates the designers' familiarity with customer requirements and a clear strategy for the NPI effort to follow. NPI targets for product quality and unit cost and commitment to a specific NPI delivery time and budget need to be identified at this time. (This subject is discussed in detail in Chapter 3.)

Issues regarding internal development of needed technology versus external acquisition are addressed. Engineering design specifications and a manufacturing plan (including a delivery schedule and an outline of work force training requirements) should be produced. To the extent that external suppliers will be involved, options are identified and their overall capability is assessed. Capital requirements are determined and the timing of acquisition for production equipment and tooling is identified. In all these areas, risks and opportunities need to be assessed in enough detail to justify moving on to the next phase of development.

Depending on the nature of the product, other types of activities will also be included in a thorough concept development stage. The product's feasibility from a patent viewpoint is explored to ensure that there are not likely to be legal barriers to proceeding. Also, customer usage patterns are projected, warranty plans are formulated, and associated costs are projected.

Since the resources required to proceed beyond this phase are large compared to those expended through this phase, it is of the utmost importance that conceptual design be a thorough process. Many companies, at this point, try to encourage considerable cross-functional collaboration. Some companies produce sketches—or even nonfunctioning models (in wood, foam, or other material)—to make the prospective product appear more tangible. The central question that one company in a retail consumer market asks at this phase is: Will developing, producing, servicing, distributing, and marketing this new product be a highly desirable use of company resources?

Specification and Design (Phase 2)

During the specification and design phase, detailed specifications for the product and the production process are worked out. Some companies call this the engineering design phase. On the product side, all questions as to what the product will do, what it will look like, and how it will be used must be addressed and answered. In appropriate situations, this is also when patent protection is sought and potential issues of environmental impacts (from either the product in use or its production process) are identified and described. The goal of this phase is to achieve a design release for the new product.

In earlier times, many companies tended to uncouple product engineering from process engineering, with the two activities occurring in sequence. The image of product engineering throwing product blueprints "over the wall" to manufacturing was an apt organizational metaphor, particularly in large organizations with complicated products. The costs of this traditional approach—in terms of unproducible designs, and unnecessary extra expense and delay due to avoidable cycles of redesign in light of manufacturing realities—became obvious in many industries. Now it is commonly accepted that product and process development belong to a single phase of the NPI process in a more integrated cross-functional work setting. Engineering prototypes of the new product are developed to establish technical feasibility and to serve as a basis for the development of a suitable manufacturing capability.

Manufacturing engineering, in the accepted mode of this phase, needs to assess product design options in terms of their ease and cost of production and implications for the resulting quality of the manufactured product. Early functioning prototypes may be developed to explore certain product or process assumptions. The outcome of such early prototypes is either that the assumptions are validated and approved for more refined development or that unanticipated issues are identified and resolved through appropriate design changes.

"Rapid prototyping" is a new technology-driven approach being followed by many manufacturers (such as cars, computers, and jet engines) to cut costs and time in bringing new products to market. This approach, which relies on computer-aided design (CAD), leads to the early identification and resolution of problems in the form and fit of assembly parts before the designs have been subject to costly drawings, tooling, and production parts. Using computer-aided engineering (CAE) and computer-aided manufacturing (CAM) integrated with CAD, rapid prototyping systems combine physics, material science, electronics, and computer graphics. Parts are produced using processes such as stereolithography, fused deposition modeling, and selective laser sintering. Both short-run and production tooling can be created through these rapid prototype systems, thereby facilitating the transition from prototype to production.

More traditionally, special processes for producing complete functioning prototypes are developed, often in a "prototype shop" in engineering or manufacturing but sometimes through outside contracts. Some industries also devise methods and tools for analyzing the results of these prototype runs. Following this "build and test" strategy, companies eventually specify and cost out a full bill of materials (BOM), including all parts and components of the product. Delivery schedules for components and subassemblies are then established and the qualification of vendors begins. Operator skills are defined and training begins.

To the extent that manufacturing engineering works closely with product designers in this phase (and probably earlier as well), savings in terms of both cost avoidance and faster time to market can be achieved. To the extent that this does not happen, costly reduction efforts (discussed later) can be expected after the product has already been introduced to the market.

Prototype Production and Testing (Phase 3)

More attention to the early phases of idea validation and conceptual design promotes product designs that will require fewer subsequent changes. The strategic importance of those first two phases, summarized in Figure 2–3, is receiving increased recognition and is discussed further throughout this book. Even when there is considerable involvement of manufacturing and customers during Phases 0, 1, and 2, subsequent additional learning inevitably takes place. The prototype testing phase of NPI offers an opportunity to learn whether the product, under realistic conditions of manufacture and use, is apt to achieve the quality specifications set for it and whether this indeed meets competitive needs. Some companies call this the product verification phase. In the case of a complex "systems" product (such as a mainframe computer), it is not uncommon for this phase to be broken into separate subphases such as engineering model build, initial assembly, integration, and acceptance. Here, major components are often tested independently, followed by a test of the integrated system.

During this phase, the complete product is produced in low-volume pilot runs and is tested under various conditions that approximate the full range of typical customer usage environments. The basic purpose is to discover any defects in design and manufacture that could be modified before volume production and shipment occur. To the extent that concep-

FIGURE 2–3
The First Two Phases of NPI

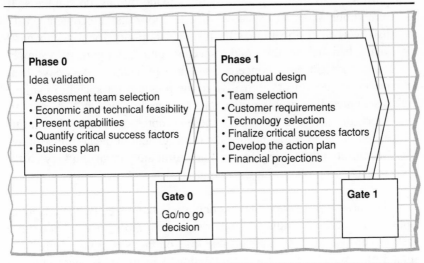

tual and detailed design have been carefully and thoroughly conducted, the costs and risks associated with prototype testing will be significantly reduced.

The goal of the prototype production and testing phase is to achieve a validated design and manufacturing release of the new product. Accordingly, it is important not only that a company do prototype testing and do it with care, but it is just as important that the right people pay attention to the outcome of such tests. As senior management demands faster product development cycle times, it is inevitable that, in some instances, the crucial prototype testing stage gets cut short. The following unfortunate example provides ample warning to others.

General Electric Compressor Example. In the mid-1980s, GE's Appliance Division in Louisville, Kentucky, was in the midst of developing a new rotary compressor for its new model of refrigerator. Seeking to leapfrog their Japanese competition, they had accepted an innovative design for the compressor, the critical component that creates cold air. A compressor prototype was built and tested in a limited fashion, in a race to get the new product to market. Although the test data looked good, flaws in the compressor design later led to field failures. Over a million of these compressors, produced in a new $120 million factory developed for this purpose, had to be replaced by units made by outside suppliers. It turned out, upon retrospective investigation, that a technician, who had worked in the Louisville testing laboratory for more than 30 years, had disassembled and inspected compressor parts on a sample basis and had been concerned about a problem of excessive heat buildup. Supervisors, sensitive to the pressure for speed to market, discounted this problem and top management was not alerted. In short, the pressure to produce these compressors quickly had influenced the test results, pushed the development of the compressor too fast through the testing process, and ultimately created a very costly and serious problem for the company.

The prototype testing phase is being revolutionized by the advent of options to conduct simulated, rather than laboratory, tests. Laboratory tests are costly and consume valuable time on the NPI critical path. Simulated prototype testing is not a total substitute for subsequent physical product tests, yet because it can be performed earlier in the development phase, it provides rapid midcourse feedback to product and process designers.

Accordingly, considerable effort has been devoted for many kinds of "high-tech" products—from flexible machining centers, to complex microprocessors—to develop computer simulation tools for rapid prototype testing. These sophisticated software tools allow the designer to "see" the results of a compressed period of use of the new product under different simulated conditions of use. Important insights on the functionality or reliability of a product still at the prototype stage can lead to cost-effective design improvements. The use of such simulated tests is often limited, however, by the underlying knowledge of the core technological processes. For copy machines, neither the electromechanical aspects of paper handling nor the marking technology is well enough understood at this time to be modeled mathematically.

Specialists in the field of industrial design, being traditionally concerned about the ease-of-use of a prototype product can also benefit from the existence of new software-based design tools. Using special software packages, industrial designers can easily assemble all of the elements of a hardware or software interface on a computer screen and test them on a sample user. This can save months of creating and testing physical models created through manual design tools.

Thorough prototype testing means that when the product is made in a volume production facility and released to the market, there are less likely to be major surprises. This is a vast improvement over efforts dominated by problems detected during volume manufacturing and subsequently when the product reached the market. Recall that in the late 1970s, Xerox Corporation was shocked to discover that a marketplace they owned rejected new products because they didn't work in the field under normal operating conditions. It took a major revolution in the way products were designed and tested at Xerox for the company to regain a competitive position in the industry they had founded. Now Xerox goes through several cycles of prototype builds, each aimed at validating the latest design. The goal is to settle on a workable design before moving to pilot production, where such changes can be costly and time consuming.

The prototype production and testing phase also has a strong market orientation in many industries. In high-technology companies, marketing usually sponsors a series of alpha and beta site tests of the new product with associated customer evaluations. Leading customers are usually used for this purpose and their feedback can influence the commercial version of the new product. Warranty requirements, which were expressed in earlier phases, need to be finalized. Detailed plans for product launch continue to be developed and training programs for the sales force and field service workers are established. Interestingly, it is only at this phase, that some companies settle on a name for the new product; up to this point many companies use an internal code name for the product.

Manufacturing Ramp-up (Phase 4)

The new product is now ready to be introduced. This final phase is sometimes called *commercialization* and includes a considerable amount of marketing activity associated with the implementation of the sales plan and the transition of product responsibility from the NPI team to the line marketing and manufacturing organization responsible for ongoing business. We limit our discussion here to the agenda within manufacturing. The central task here is to gradually achieve the manufacturing capacity (commonly called *ramp-up*) necessary to meet the projected sales volumes, while successfully meeting the new product's targets for unit cost, conformance to performance specifications, and other measures of quality, including customer satisfaction.

The ramp-up activity can be very challenging for manufacturing. Action should already have been taken to assure that all necessary equipment, tooling, and other processes will be available at the right levels of capacity. In the ramp-up phase, the work force must be adequately trained and supervised as they begin to produce the new product at these commercial levels. Successful ramp-up also requires ongoing supplier management to assure the consistency and timeliness of incoming materials and parts. In some industries special effort is required to ensure that the manufacturing process itself is operating as designed. This often requires the intentional pursuit of improved product quality and cost by reducing the variability of manufacturing processes. Skills in optimizing production process plans and learning from ongoing production experience are particularly critical at this time. In addition, the manufacturing organization needs to build the capability to support field service requirements as they emerge.

During the ramp-up stage, key members of the product design team will still be involved. Their skills are often needed in efforts to optimize the manufacturing process. In addition, as problems are identified in achieving cost and quality conformance, redesign of the product may be required. Companies that have employed cross-functional teams in the earlier phases of conceptual design, product/process development, and prototype testing report reduced requirements for product redesign during ramp-up. Even so, some activity along these lines must be expected.

As the new product enters the commercial realm, some companies place considerable emphasis on customer acceptance. Aftersales activities along these lines include installation, possibly through field introduction teams. Activities by marketing, customer service, and field service may be considerable at this time (some of which is discussed in Chapter 12).

Cost Reduction: A Common Follow-up Stage

In some industries, manufacturing ramp-up is not the final stage of new product design and development. Whenever speed-to-market and/or product quality are taken to be top priorities, there is a significant chance that the element of unit product cost will be treated at first with less rigor. The common result is that initial product costs exceed their objective.

At this point, companies often engage in formal cost reduction efforts. An example of this occurrence, at GE Aircraft Engines, is described in Chapter 10. In this example, the issues associated with cost reduction were tied primarily to the resulting engineering costs associated with product design changes, new hardware costs associated with requalifying the engine and its parts, and additional tooling costs.

Participation in such a cost reduction effort should include key individuals with expertise in manufacturing, materials procurement, finance, engineering, and marketing. Their primary responsibilities would probably include the following:

1. **Manufacturing.** A revised manufacturing plan including make/buy improvements, a review of the cost and performance characteristics of tooling and other hardware, assessments of the application of new manufacturing technology, and a review of cost estimating techniques. This effort would include an estimate of the productivity gains to be expected from the recommended changes to manufacturing procedures, process flow, and operating parameters.

2. **Purchasing.** Analysis of the cost of the last materials purchase, the current one, and the planned next one; consideration of implications for the overall product cost of changing materials prices at time of projected delivery; examination of cost trade-offs for placing larger quantity purchase orders for needed materials.

3. **Finance/Cost Accounting.** Development of learning curves for each production shop; review of cost aggregation and overhead allocation techniques; definition of inflation factors to be used in cost forecasting; investigation of shop managers' responsibility for managing raw material and overhead costs.

4. **Engineering.** Review of trade-offs between cost and key product characteristics (e.g., weight or size) in search for acceptable ways to reduce costs; comparison of the manufacturability of this product with earlier designs looking for additional design improvements; review of earlier cost reduction options that were denied on engineering grounds. This effort would include the preparation of a plan, including the associated timing, for achieving the original product cost objectives.

Cost reduction efforts of this type are an expensive and time-consuming use of resources. The only way to avoid such efforts is to design the new product to achieve the desired level of unit cost at the initiation of volume manufacturing. This concept, often called *design for mature first cost*, is a challenge to implement and is discussed in Chapter 3.

Also note that even without a major cost reduction program, products and their production processes are improved as part of ongoing efforts by design engineers, manufacturing engineers, and other members of the manufacturing organization. Such product improvement activity is often originated by ideas from the field service or marketing organization or directly from customers themselves. In fact, even though the NPI team may have been disbanded long ago, such product improvement activity may be considered to be normal and ends only when the product loses its market or is made obsolete by the introduction of a new product.

GATE REVIEWS

Many companies have formalized their NPI process to the extent that they identify the various phases of work and the executive review processes that occur between any two consecutive phases. A commonly used term for these reviews is *gate*. The term is intentionally graphic. A special committee composed of senior managers from all business functions reviews the progress of a product design and development project at the completion of a specified milestone in the project. Each gate review calls for well-specified deliverables. If all is well at this point, the project officially passes through the administrative gate to the succeeding phase of work.

In practice, however, the phase-gate system in many companies is not followed as closely as their detailed and precise operating manuals might suggest. In particular, the pressure for increased speed to market for new products has been a catalyst for considerable overlapping of the phases. Most notable is the growing acceptance of "simultaneous engineering" (discussed later in this chapter) in which some tasks logically associated with one phase are allowed to begin in prior phases in order to save time on the critical path of the overall project.

Each gate involves a formal review of the status of the NPI project compared to its original targets for the product's functionality, quality, and cost. Under this event-driven control system, preestablished checklists are followed to examine the latest set of product and process development decisions, the success in implementing prior decisions, and the anticipation of possible downstream problems. If prespecified exit criteria are met, the executive review is successfully concluded, and additional resources are allocated to the project and various responsibilities assigned for the next phase. If, however, a project fails at any one of these reviews, the project is either canceled or it remains in the current phase until the identified problems are remedied and that particular gate review is passed. The most difficult managerial challenge in gate reviews is to establish clear exit criteria and to be unequivocal in applying them.

Sometimes a gate review becomes the occasion for transferring formal leadership of the project to someone new, reflecting the priority challenges to be met at the upcoming phase. In an extreme case, the leadership may pass from someone strong in marketing at the earliest phase, to someone skilled in engineering management during product development and testing, and perhaps later to someone with a strong track record in manufacturing or marketing depending on the most complex aspect of the product launch. Even during the technical design phases of a project (as illustrated by the case of the Norstar product in Chapter 8), the leadership may pass from one engineering manager to another, based on the skills and styles needed at the time. Clearly, senior management must assess the trade-offs involved in such shifts in project leadership, based on an appreciation for the talents of their project managers and the shifting needs of their NPI projects. All other things being equal, continuity in project leadership can be advantageous.

The basic objective of the gate review process, then, is to keep NPI projects on track, while making risks and trade-offs explicit. This formal review process intentionally encourages cross-functional problem solving throughout the NPI cycle. It also provides important feedback to top management, thus helping to sustain their involvement and commitment throughout the entire cycle.

Companies that follow the formal set of phases outlined above, are likely to adopt the associated (and similarly numbered) set of gate reviews described below. You should understand, however, that this description of gate reviews, as with the prior summary of phase activities, is provided as an extended illustration. It is not meant to be a cookbook proposal for the best system to follow. Nor is it meant to be a fully comprehensive discussion of everything that occurs within any particular company.

Launch the Project (Gate 0)

The purpose of this initial gate review is to establish the economic and technical feasibility of investing company resources on the proposed new product ideas. This review would be a much more detailed exercise compared to what would take place earlier when the new product idea was originally screened. By now considerable research has been done on that original idea. The executive review committee is now in a position to decide whether to proceed through the conceptual design phase, to wait until particular questions are resolved, or to reject the proposed new product and cease all further design and development work. Companies that conduct rigorous Gate 0 reviews seem to agree that this is the most strategic point for managerial intervention. Indeed the labeling of this gate signifies its importance; companies that began their executive review with Gate 1 discovered that they were entering too late into discussions of critical formative issues of the market and performance for the new product. They subsequently added Gate 0 to the front end of the preexisting set of gate reviews.

Gate 0 assesses the full range of questions about the market need for the proposed product and its strategic fit with the company. Initial NPI targets, such as the time of the first customer shipment, development cost, and measures of quality for the new product as well as its projected unit cost and price elasticity are presented. Issues of technical challenges and risks, investment requirements, likely profitability, impact on current operations, relative priority, and availability of resources are also pursued. The plan to demonstrate feasibility in the conceptual design phase is also reviewed in detail. Successful completion of the Gate 0 review results in the allocation of funds to conduct conceptual design, the identification of a program (or project) manager, and approval to assemble the needed human resources.

Approve Project Implementation (Gate 1)

At this gate review, management reviews the full project plan for achieving the objectives of the NPI. Central to this review are the definition of the customer/market, the business plan, and the feasibility of the project from a combined technology, manufacturing, and market point of view. This review also includes detailed product specifications and the schedule and other plans for executing the NPI project. At this review, management must feel comfortable that the designers' approaches accommodate manufacturing realities. Some companies also consider at this time factors such as patent potential, and projected environmental, health, and safety impacts.

The Gate 1 review also includes an assessment of the proposed new product effort with other ongoing programs and priorities. In some companies, new products are few and far between, and each of these clearly merits a top priority designation. Other companies introduce many new products and must therefore constantly review the priorities within this portfolio. The status of ongoing business can also strongly affect the priority given to a new product effort, even if its potential, as assessed in the Gate 1 review, appears to be high.

Successful completion of the Gate 1 review results in a decision to proceed through the design phase and the resources required to conduct this phase would then be allocated. If there was insufficient information presented to justify this decision, the team would be asked to address the points in question and to return for a rescheduled Gate 1 review. Other options are to suspend the program pending improved climate sometime in the future or simply to terminate the project. An improved climate might refer to market conditions (greater potential sales or profit margins), financial conditions (more funds available to invest in such a product), or technological conditions (more fully developed and readily available capabilities).

Release the Design (Gate 2)

The Gate 2 review is aimed at achieving a design release of the new product. This review often requires a prototype demonstration of the complete product and the availability of critical process capabilities necessary to produce it. Executive reviewers seek to confirm that the product design meets the specified requirements (i.e., functionality, weight, and size), that its delivery can be executed in accordance with the business plan (such as product availability and unit cost), that there are no major unresolved issues of technical performance, and that the resources and time still required are within the intended bounds. At this time, marketing may be called upon to present evidence that the customer response to the new product is positive.

Begin Volume Manufacturing (Gate 3)

The Gate 3 review is aimed at providing approval for manufacturing ramp-up and market release of the product. An important consideration at this time is whether all aspects of the manufacturing capability are ready for initiating the gradual acceleration (ramp-up) to full-scale production.

The team must demonstrate that the previously established criteria for product quality have been met and that the marketing preparation for launching the product has been completed. The outcomes of a successful Gate 3 review are: a confirmed product launch date and, if necessary, approved funding for any cost improvements or design modifications.

THE PRACTICE OF SIMULTANEOUS ENGINEERING

Many manufacturers have been stunned in recent years to discover that their competitors were developing new products in half the time and at half the cost that they had been experiencing. One of the key ingredients to such performance success is the switch from sequential to "simultaneous" engineering of new products and the manufacturing processes to produce them. The difference between the two approaches is illustrated in Figure 2–4.

Sports analogies have been widely used to distinguish between these two approaches. The more traditional sequential approach is portrayed as a relay race, in which only one player runs at a time and a baton is passed in one direction from one runner to his or her successor. If any one runner falters, or if the handoff is bobbled, the entire effort is delayed. Simultaneous engineering, in contrast, is likened to a game of rugby. Here, the entire team runs down the field at the same time, repeatedly passing the ball back and forth among the players. Since the two sports events are different, the analogies that can be drawn between them are limited, but the imagery is clear: Simultaneous engineering (like rugby) requires considerable ongoing interaction and is at all times a team effort.

In many companies, terms such as *simultaneous engineering*—or other labels such as *concurrent engineering* or *codevelopment*—have come to be the loose equivalent of integrated, cross-functional teams (including suppliers and customers). Chapters 3–5 deal with various aspects of team integration throughout the product design and development process. Here we concentrate on the original core notion of simultaneous engineering involving internal engineering groups of the company.

The Original and Narrow View

The original and narrow view of simultaneous engineering is simple but powerful; manufacturing engineers are brought together with product de-

FIGURE 2–4
Sequential Approach versus Simultaneous Engineering

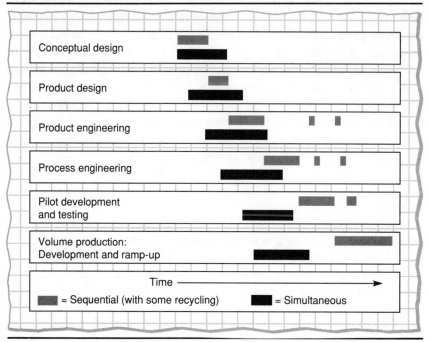

sign engineers with a combined objective of developing better products than would emerge from their traditional modes of partial isolation from each other. In this new mode of interaction, manufacturing engineers get involved early when the product concept is being refined and its design is beginning. The manufacturing engineers then have an opportunity to ask questions about product design and present existing manufacturing capabilities and capacities before much time or effort has been put into any single design alternative. In this manner, aspects of product cost and aspects of quality affected through manufacturing can be considered in more detail sooner than would otherwise occur. This approach thus saves the extra time and expense that would be incurred for product redesign under the sequential, "over-the-wall" approach in which manufacturing engineers are simply asked to develop a process for producing the new product as already designed. Through integrated thinking, in which design engineering considers additional requirements (e.g., manufacturing, cost, procurement cycles) besides the desired functionality and performance of the new product, these benefits can be achieved.

Even within this narrow view, simultaneous engineering means more than integrated thinking about the design of a product. It also means starting certain engineering tasks related to the manufacturing process at the same time that details of the product design are being worked out. In other words, there is codevelopment of product and manufacturing process, as work in these areas overlaps in time rather than following a strict sequence of upstream product decisions leading to subsequent downstream process decisions. For example, the design and procurement of critical tooling may begin before there has been a final sign-off on the design of the part for which the tooling is to be used.

Benefits

Simultaneous engineering is one organizational response to the need for an NPI work structure that is both more efficient and more effective. This approach is consistent with the pursuit of several important capabilities (discussed further in this chapter).

Discipline in the NPI process is strengthened when design and manufacturing engineers are assigned to work together (often in the same physical location) with a joint responsibility for producing effective product and process decisions.

Enhanced identification and reduction of risk is a direct outcome of the shift to simultaneous engineering because more issues of technical feasibility are raised early on. In this manner a project manager or an executive review team are more able to spot priorities for additional problem solving at a point where the cost (and time) to solve these problems is comparatively small.

Problem resolution is strengthened as people with the requisite skills and experience interact with each other in an ongoing way with a common objective, rather than occasionally and with conflicting objectives.

Resource allocation is more efficient because a company can expect the early view toward manufacturing (inherent in the pattern of overlapping problem solving that occurs under the simultaneous engineering approach) to require fewer aggregate engineering hours. The fact that shorter development cycle times can be accompanied by increased productivity has been demonstrated in a number of settings, most notably in research on the world auto industry by Kim Clark and Takahiro Fujimoto (1991, chap. 4).

Simultaneous engineering promotes the achievement of loose control that is so beneficial to effective new product development. When design and manufacturing engineers work together, it is less likely that disagreements will be escalated to higher levels of management. Accordingly, executive gate reviews under modes of simultaneous engineering will concentrate more on the nature of the solutions that were jointly developed rather than on reenacting traditional functional "turf battles" and adopting heavy-handed forms of management control.

Potential Problems

Despite these advantages, simultaneous engineering engenders its own set of potential problems. Management at all levels must be aware of these if they are to have a realistic understanding of what to expect from this kind of work structure:

1. Simply bringing people together with a common purpose does not guarantee that they will function as a coherent team. Boundaries that have grown between design and manufacturing departments through the years cannot be expected to disappear at once. This issue is discussed in detail in Chapter 4.
2. The evolution of CAD/CAM/CAE tools has made it possible to accelerate the pace and effectiveness of simultaneous engineering efforts. However, such design-based technologies require discipline in establishing centralized, standardized data bases and appropriate training of all who use them. Issues in the effective use of design technologies are discussed in Chapter 6.
3. Simultaneous engineering, where some downstream activities start before the upstream activities that shape them are finished, may require extra communication and flexibility. Manufacturing engineering, for example, may place an early order for certain tooling based on an emerging product design only to be told that the design needed to be changed and that the tooling needs to be modified. Clearly, in such a situation, it is important that manu-

facturing engineering know about the change as soon as possible and that the tooling source (either inside or outside the company) be selected with flexibility in mind.

A Broader View

Simultaneous engineering has taken on broader meanings as companies have gained experience in bringing together more sources of expertise and knowledge during the product design and development process. In this broader view, customer involvement throughout the process of design, development, and testing is seen as central, as suggested by the earlier descriptions of the various phases of work. The point here is clear: customer satisfaction, which is essential to NPI success, can be affected by design and development choices throughout the project. Whenever a customer orientation might be lost without customer involvement, companies are seeking effective means of introducing this involvement.

The early and ongoing involvement of primary suppliers, responsible for the design and manufacture of important parts or subsystems, is also part of the extended practice of simultaneous engineering. Forming strategic partnerships with suppliers—sharing with them the goals of the NPI, and having them play a role in conceptual design, project planning and scheduling, and even working jointly on key design and development activities—is increasingly common. The strategic inclusion of suppliers as part of a simultaneous engineering approach can improve customer satisfaction and cut the time to market.

Early and ongoing involvement by company experts in field service, cost accounting, and purchasing have similar benefits. The early involvement promotes simultaneous consideration of multiple issues during concept investigation. The ongoing involvement assures that the design of the product, the manufacturing process and required tooling, and the selection of suppliers of raw materials and components are compatible with the original project objectives. For experts whose skills are needed only occasionally or those whose activity is on the critical path of the overall project, the timing of involvement is critical.

A Comment on Process Industries

The practice of simultaneous engineering—and indeed the whole model of phase reviews for new products—applies to products that are fabricated or assembled in small or large lots, as distinguished from continuous process industries, such as chemicals, pharmaceuticals, paper, or ceramics. While this book does not attempt to deal with the problems of product design and development in specific industries, a brief summary of what makes process industries special is appropriate at this point.

When a customer request or market survey in these "process industries" signals the need for a new product, the limits of existing production processes may be too constraining and a new process will have to be developed. The design and development of such products is so inherently tied to the process through which it will be made that the emphasis of new product introduction in these settings is on the development of the production process. One initial activity is to design the production equipment (which would include pumps, pipes, and reactors for producing chemicals, and injection molding machines and furnaces for ceramics). Companies in these industries then have to deal with the inherent uncertainty in the new production process; what it can produce and how this relates to the settings on the equipment.

Because different settings determine the properties of the resultant product, the production process, in a sense, *is* the product. A new product effort, to be successful, must then identify properties of the product that will meet customers' application needs, while being technically feasible to achieve. Typically, the balance is reached through cycles of experimentation.

Establishing the properties of materials that can be produced requires experimentation under highly controlled laboratory conditions by the research arm of the organization. Once the desired product is produced in the lab it needs to be scaled up for commercial levels of production. New problems of feasibility are then inclined to arise, since the commercial version (designed for a much higher capacity) will not be a replica of the laboratory version, and the associated processes are often likely to behave differently. At both laboratory and commercial levels of production, test engineers try to ascertain how the operating parameters of the process, such as time, temperature, and pressure, will affect the properties of the resulting product. In industries making materials for use in parts of end products, the new material also must be tested by the customer, as part of the overall development effort, to see if it performs well in the intended applications.

These are some of the realities of product design and development in continuous process industries. While there are similarities with products that are fabricated or assembled, there are also significant differences. There is no generally accepted formal structure in these industries that is equivalent to the phase-gate system of review as described in this chapter.

MANAGING THE PHASE-GATE PROCESS

It should be clear from the above description that the phase-gate approach to product design and development may serve several important functions:

- Achieving discipline and consistency.
- Identifying and reducing risk.
- Allocating resources.
- Achieving "loose" control.
- Empowering the team.
- Resolving design problems.
- Smoothly shifting from development to volume manufacturing.

Each of these functions is briefly discussed below.

Achieving Discipline and Consistency

New product development is a complicated process, particularly when the company is sizable and the product is somewhat complex. The phase-gate approach was developed to facilitate consistency of this process. It accomplishes this by specifying what needs to be done in each phase and how the work in each phase will be assessed in the subsequent gate review. Companies that adopt the phase-gate approach usually specify the various kinds of documents that must be prepared as a basis for each gate review. Although the implementation of this process of formal review has its own cost and time requirements, its widespread adoption signals the belief that the benefits are worth it.

Identifying and Reducing Risk

With respect to both technology and market dimensions, the phase-gate approach is designed to identify risk in advance and to monitor that risk as the project proceeds. Naturally, no administrative process alone can reduce risk for this is the job of people who are on the NPI teams. In the case of the GE refrigerator compressor, reviewed above, the underappreciation for the element of risk spelled ultimate disaster. Because the company had already made 12 million rotary compressors for air conditioners, management assumed that to design and manufacture this type of a compressor for refrigerators would be a routine exercise. Because the inherent risk had been underestimated, the Appliance Division's chief design engineer rejected a proposal to develop the rotary compressor through a joint venture with a Japanese manufacturer already experienced in making a refrigerator rotary. Similarly, the GE team rejected the option of using as a consultant the engineer who had designed the original air conditioner rotary compressor at GE several decades earlier. Management did not employ these risk reduction devices because they did not perceive the level of risk that was inherent in this project.

How can such misperceptions be avoided? One way is to encourage, in fact to require, the early identification of risk. This is facilitated (but not guaranteed) by using gate reviews as formal occasions to ask questions about project risk and how it will be handled. At this time, the project team needs to be willing to deal objectively with matters of risk, without fear of retribution.

Allocating Resources

Resource allocation requirements will vary throughout the phases of an NPI project. As previously described, one of the primary activities in early phases of any NPI is to estimate the resource requirements for the remainder of the project. Gate reviews provide a formal opportunity to review the efficacy of resources already allocated and to clarify the need for upcoming resources. On an ongoing basis the project (or program) manager will assess whether existing resources are being well used and if any modifications from existing plans and schedules will be required. The one lesson that seems to be common across industries is that it is worth allocating serious levels of resources (and a team of people with all of the relevant functional backgrounds) to the work activity preceding the Gate 1 review. As will be discussed in Chapter 3, this turns out to be a very effective investment, reaping benefits in the total time to market as well as to the cost and effectiveness of the product itself.

Achieving "Loose" Control

The gate review process is clearly a form of management control. Senior executives from all relevant business functions have an opportunity, at these event-driven occasions during product design and development, to exercise control on the future of that project. However, the gate review system clearly works best when those team members who are familiar with the proposed product, production process, and market, try to anticipate and resolve problems themselves. The intent of this design and development process is not fulfilled when senior management intercedes, late in the process, with their own suggestions for product design changes. An effectively implemented phase-gate system can serve to discipline senior management as well as project team members: it can set bounds on what questions management will ask and when they will ask them. This type of intentionally limited and disciplined management control seems to characterize the companies that are most successful in fast and effective new product development.

Empowering the Team

Through concentrating its involvement early in the project, senior management can maximize its own impact while empowering the product team. Senior managers often show little interest in a project until it is in the prototype phase; at this late point, they can have little impact without upsetting the entire project, and their intervention is experienced as withholding power from the team. If, on the other hand, senior managers concentrate their attention on activities during what we have called *Phases 0 and 1*, they can have enormous impact, yet still empower the team by keeping out of decision making during implementation. Figure 2–5 depicts the inconsistency between the late intervention that seems still to be common in many companies and the ability to influence the outcome.

Empowering an NPI team, therefore, does not mean that senior management "gives up" control. Rather, it refers to the assumption of responsibility in matters of overall strategy, and the relinquishing of responsibility for the execution of product design and development. If senior management has done its job well (i.e., establishing capabilities, strategy, targets, and review processes), then the product development team is clear about the parameters within which it can act, and the resources upon which it can draw.

FIGURE 2–5
The Need for Management Attention in Early Phases

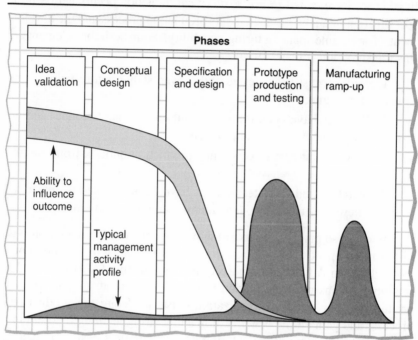

Resolving Design Problems

Product design is a process in which options are identified and analyzed. Given the multiple objectives of any given NPI, one can expect important trade-off issues to arise as different design options are considered. Problems will arise in seeking to satisfy these various objectives. Many of these problems will be resolved by the members of the NPI team. Indeed the purpose of the phase-gate system of new product development is to encourage the ongoing design team to anticipate the questions to be raised at the upcoming gate review and to try to resolve all of the problems that are scheduled to be exposed at that later date. To the extent that the team has not identified some problems or has not fully resolved the ones that they had identified, the gate reviews will force the necessary analysis or decisions to occur. The program (or project) manager plays a very significant role in the problem resolution process, as will be described in Chapter 4.

Smoothly Shifting from Development to Volume Manufacturing

In what we have called *Phase 3*, a critical transition occurs between the specification of a new product and its production process and the development of a volume manufacturing capability. Four desirable elements of this transition are (Ebner, 1990):

1. New processes are tested and proven stable before volume production begins.
2. Workers have been trained in building the new product and are achieving the desired quality levels.
3. Mature first-cost targets guide production outlook from the outset.
4. Vendors and their technology have been seamlessly integrated into new production through careful planning and timing.

Because so much is at stake in moving from a new product prototype to the ramp-up of a full-volume manufacturing capability, some companies have an Advanced Manufacturing Engineering (AME) group that sets up and runs pilot production processes well in advance of volume production. This interim organization exists to debug the product and process, and gain all the benefits of manufacturing learning before transfer to the full-volume manufacturing site. It identifies design problems, develops and tests manufacturing processes, tools, and fixtures, and trains the work force.

Methods for executing this transition vary. Here are two typical examples.

At AGFA Compugraphic, a worldwide leader in the design and manufacture of electronic prepress equipment, the AME organization produces 100 to 300 units over a three- to five-month period (Tatikonda and Rosenthal, 1990). This follows a design engineering prototype period, in which 10 to 20 systems are built by AME and design engineering working very closely together. AME prototype production looks at design process refinements (including safety concerns), manufacturability, serviceability, testability, and performance certifications. Product design changes at this point are hoped to be minimal, but they are made as necessary. The team includes all of AME, manufacturing and test engineers involved in tooling development, and most of the design engineers for the product. They work further on the production process, keeping extensive logs, using production tooling and parts as they become available, with daily problem-solving meetings on the shop floor. Risk assessments are done, contingencies pursued, and "what ifs" explored. An advance guard of employees are trained, and then transferred back to the volume production facility, along with the product, assembly equipment, tooling, and technicians.

At Amdahl Corporation, a price/performance leader in the design and manufacture of large mainframe computer systems, AME builds the engineering models in a prototype area called the Initial Production Organization (IPO) (Venugopal, 1990). Design problems that are uncovered go back to design engineering. AME develops and tests manufacturing processes, proves out new equipment, and builds five or six systems. A very skilled work force staffs the IPO, and works with the advanced manufacturing engineers on developing the detailed processes. They become the lead people in the production area, and train additional personnel for the ramp-up. IPO takes place in the period 18–24 months prior to the first customer shipment. During this time, manufacturing shares the lead with engineering and product management for building the product, but has the major role in seeing that it gets done on a timely basis. It is not unusual at this stage to identify some aspect of the product that needs to be redesigned. This often requires the "remake" of custom chips in which case engineering sends appropriate information to the chip vendors. Fast turnaround time from vendors is necessary and is negotiated at the beginning of the NPI, along with associated cost penalties for late delivery. During this period and thereafter, manufacturing is also installing equipment, training people, and preparing for the ramp-up.

Different companies will have different approaches to managing the development/production interface. The question is whether the chosen approach is effective in achieving a smooth transition from the product development team to the manufacturing organization responsible for routinely shipping high-quality new products at the required volume levels.

SUMMARY

Established companies usually define their NPI process in terms of several distinct phases or stages. This is done to encourage a consistent and disciplined approach to product design and development. Consistency is achieved by raising and answering a standard set of questions at pre-scribed points in the creation of any new product. Discipline among NPI participants is required to anticipate these questions and to justify design concepts and specifications in a particular sequence. Discipline by senior managers is required in objectively reviewing the prescribed deliverables and then opening the gate to the next phase *only* when all major issues have been resolved. It is important to understand how the definition and management of the phases and gates serves to structure the work of prod-uct design and development and, therefore, shape its outcome.

Many companies, faced with competitive pressure to introduce bet-ter products faster, are deeply involved in examining their existing struc-ture for designing and developing products. Two approaches currently dominate company efforts to change the structure of work toward more effective product design and development. The first, relative weighting, increases the emphasis on (and resources allocated to) early phases of the project. The second acknowledges the interrelationships of the various phases of work and the importance of compressing and overlapping such work.

More and more companies are recognizing the critical importance of a thorough product concept and project plan preceding decisions on the design and development of the product. As a result, the early phases are claiming a larger fraction of the time and attention that management gives to new product introduction. One indicator of this shift in emphasis is the adoption of language and structures that give this early work a special status. The term *Phase 0,* as used in this chapter (and in companies such as General Motors, Northern Telecom, and Hewlett-Packard), signifies the importance of work to validate the ideas for the new product before the NPI project formally begins. These ideas are further refined and de-veloped as part of Phase 1. The importance of these early phases is cap-tured, in the words of one senior manager in a leading high-technology company: "If we don't manage Phases 0 and 1, we don't get there. As someone who has spent many years working as an engineer, it pains me to say this, but the rest is boilerplate."

Despite the orderly nomenclature for the phases, several iterations are often required within a phase and different functional tasks (e.g., marketing, product engineering, and manufacturing) will occur simul-taneously. While new product development is often described as a set of functional tasks occurring in a roughly sequential fashion, the process typically lacks such linearity. In recent years, companies have been redefining their NPI phases to promote the practice of simultaneous engineering. This practice is intended to shorten the overall time to market and to generate better designs for the product and the manufacturing process. Simultaneous engineering adds to the complexity of ongoing information flow and causes design and development phases to overlap.

Prototypes provide critical opportunities for testing product concepts and detecting basic design flaws as early as possible. Identifying, diagnosing, and removing such flaws before consuming unnecessary resources and time is key to cutting lead times and increasing customer satisfaction. Testing prototypes of individual assemblies—and then the entire product—will signal remaining design and development priorities. Although early prototype development and testing can be costly, a series of short, targeted prototype cycles can have a huge payoff through reduced effort in pilot production and manufacturing ramp-up and less need for subsequent engineering change orders.

The requirements of a formal work structure to manage the entire NPI process have an effect on people in all areas of a company, as well as its key suppliers and customers. Putting this structure in place in a way that encourages rather than cripples effective product design and development is no easy matter. In the next chapter, we look within this formal structure to the planning and management of the actual NPI projects. There (and in subsequent chapters) we will see why product design and development is rarely a streamlined, efficient process.

CHAPTER 3

PLANNING AND MANAGING THE PROJECTS*

Only in the most abstract sense do formal descriptions of a product delivery process and associated controls explain what is involved in product design and development. Such descriptions, as summarized in Chapter 2, provide an overall appreciation for interim progress in the design and development of a new product. What they miss, however, is an operational sense of the work involved in design and development and *how* this work gets planned and accomplished.

Any product design and development project should begin with a plan tied to specific objectives for that project. This requires considering (at first in Phase 0 and then in a more refined manner in Phase 1) the trade-offs inherent in any new product concept. A common starting point is to try to balance elements of product performance with the resulting unit cost of the product. This can be especially difficult when considerations of conformance quality (achieving production specifications) add issues of manufacturability into product design. Broadening the definition of quality, to go beyond matters of conformance and include other dimensions important to the customer, expands further the range of trade-offs to be considered. More recently, managers have taken on the additional challenge of including options to save time: designing and developing products much more quickly without compromising performance, quality, and cost objectives. Defining these critical success factors and considering associated trade-offs and priorities in setting targets for them is, therefore, an important aspect of NPI project planning.

Project planning, however, is only the beginning because, even with the best of plans, the execution of NPI projects is rarely routine. Problems inevitably arise as resources or time constraints become apparent; unanticipated external events occur; or technical solutions fail to match a

---

*Some of the material in this chapter first appeared in a working paper by Stephen R. Rosenthal and Mohan V. Tatikonda (1990).

critical success factor for the new product. Most products are so complex that design and development activities need to be performed by teams of people with complementary sets of expertise. Coordination of the work of these experts then becomes an important ingredient of successful product design and development. Ongoing project management, as distinguished from occasional executive reviews, then takes on special significance.

This chapter examines product design and development from the perspective of project planning and management and addresses the following questions:

- What common targets should guide all those involved with a particular NPI project?
- How should work be scheduled and resources allocated to best meet these common targets?
- What is the role of the project manager?

Naturally, these questions will have different answers for different companies as they prepare to launch specific projects. Differences will emerge even within a single operating unit of any company, in response to shifts in their competitive, market, and technological environment. This chapter explains the significance of each of these three questions and identifies and explores the related project management challenges. With this orientation, managers are better prepared to provide precise answers to these three questions under actual and varying circumstances.

THE NEED FOR MULTIPLE TARGETS

A bit of historical background explains how manufacturing companies need to view the effectiveness of a product development project. It all centers on what it takes to exceed customer expectations and achieve industry leadership.

By the mid-1980s, quality had moved to center stage; and in the United States, most manufacturers are still in some phase of improving quality.[1] Some companies became stuck on defining quality primarily or solely in terms of manufacturing conformance, and some believe that

---

[1] Readers interested in statistics that track the importance of quality related to other manufacturing initiatives should refer to the findings of the series of North American Manufacturing Futures Surveys conducted by the Boston University Manufacturing Roundtable.

they've "done it" (i.e., closed the quality gap, and thus, in their minds, the competitive gap). Other firms are expanding their consciousness of quality to mean more than manufacturing conformance. By including notions of maintainability, ease-of-use, reliability, and other factors, "quality" flows into "customer satisfaction," which is deemed the critical measure of performance in the 90s.

As suggested in Chapter 1, customer satisfaction with new products requires a great deal more than quality of conformance. Global competition, fragmenting markets, rapid technological changes, as well as growing sophistication and expectations of customers, requires an even higher level of performance in these areas: speed to market; responsiveness to customers; design elegance and integrity; and flexibility and cost-effectiveness in small-lot manufacturing.

It is clear that companies must get faster—in other words, that NPI cycle time must get shorter. The challenge is to do this while avoiding the traps of oversimplification and single-mindedness that occurred with quality. Competing through time is much broader than simply achieving speed in new product development, just as quality is much broader than conformance. And, while it is clear that most companies need to get faster (often much, much faster) in developing their new products, statements of time urgency should consider factors of product performance quality and cost.

The issue then, is how do companies get much faster at NPI without compromising other objectives, and without falling into some big traps? The central challenges in setting targets and balancing their priorities are threefold:

1. To establish mutually compatible time, performance, quality, and cost targets based on external and internal requirements and capabilities.
2. To communicate a consistent message about these targets, their priorities and their interrelationships.
3. To manage the NPI process to achieve the set of targets, and, in the face of slippage of one type or another, to know which targets to reset and at what levels.

In general, four strategic targets will provide the necessary common vision for a product design and development project: development cycle time; development cost; dimensions of quality; and unit cost. The first two of these targets relate to the design and development process, while the second two relate to its output. These types of targets and their significance as success factors are described below.

Development Cycle Time

The development cycle time is generally considered to be the elapsed time from initial product concept until the new product is commercially available. The period of interest is shown in Figure 3–1. A shorter cycle time, other things being equal, raises the competitive value of the new product. Getting a new product to market in advance of the competition has the obvious advantage of capturing additional sales and higher profit margins, that is, assuming the product meets a need not otherwise met by existing products. Competitive and market concerns increasingly force companies to set more aggressive development cycle times.

A shorter development cycle time also has the not so obvious advantage of letting the company wait longer before starting the project, given a particular target date for the new product introduction. This later start date often provides opportunities to adopt newly available, better technology or to develop more timely—and therefore more accurate—projec-

FIGURE 3–1
Product Development Cycle Time

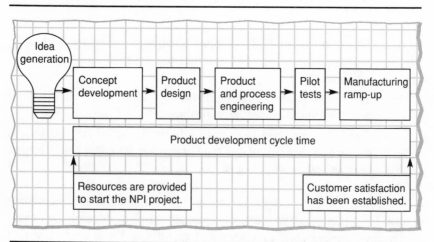

tions of customer requirements. This also reduces the company's risk of simultaneously trying to develop technology and the product that will use it.

Product development cycle time has different meanings in different companies. In fact, many companies use various terms for this general concept, including *lead time, time-to-market*, and *new product delivery time*. We will use a rather precise definition for " product development cycle time": The period that elapses from formal allocation of resources to the project until the new product's commercial delivery from volume production is demonstrated to be satisfying its customers. We explain later how this definition ties to our description of the NPI process in the previous chapter.

In practice, both the starting and ending points of this period may be subject to some ambiguity. In some companies, a product design and development project starts informally, well before resources are officially assigned to such an endeavor. This practice, often called *bootstrapping* resources, means that the official product development starting point understates the actual amount of elapsed time to move a product through the early stages of concept development. In theory, it would make sense to include the preproject Phase 0 time as the initial segment of development cycle time. Companies that allow months, or even years, to pass before coming to a decision on whether to start an NPI project are reducing the likelihood of its ultimate timely delivery. However, companies tend not to start the clock until after the idea generation phase of the NPI project is officially underway (i.e., Gate 0 is passed). At the back end, some companies will rush to market, delivering initial quantities of the new product before the volume production capability has been fully developed. This practice, according to our definition, also understates the actual new product development cycle. Our definition may turn out to overstate the actual new product development cycle, as when delays arise between the official allocation of resources and the actual start of work on the project.

These definitional ambiguities have direct significance to senior managers charged with the responsibility to make their own company's development cycle time comparable to or better than the best in their industry. Because we are always trapped into thinking in terms of available measures of performance, it is easy to misinterpret what the competition is capable of doing. As the business press continues to emphasize the importance of time-based management, one can expect a continuing flow of comparative data on development cycle times in different industries.

But exactly what time interval is being measured by such statistics? And are these statistics across companies really comparable? These are some of the basic questions that managers should ask before jumping to conclusions.

Development Cost

Development cost, another measure of the product design and development process, has two important dimensions: (1) total financial requirements and (2) associated human resources that will be needed to complete the project. With respect to funding levels, companies have to allocate the necessary funds for all phases of design and development of the new product and its manufacturing process. In most organizations of any reasonable size, this factor must be considered in light of budget realities and the timing of budgetary commitments. Second, even if the budget seems feasible, are the key human resources likely to be available at the desired levels and mixes? This question clearly raises issues of the relative priority of the contemplated project compared with other new product efforts and the production of existing products.

Even though the full set of resource requirements is not likely to be known until the project is well under way, rough targets in terms of anticipated development costs are usually requested at the beginning of the project. It is also important to plan the allocation of the needed resources throughout the life of the project. As described in Chapter 2, concept development (Phase 1) is a critical, and often understated part, of a project. Establishing the cost and resource allocation implications of this phase should be done with particular care, even though it will be a relatively small portion of the total product and process development cost. Management must be convinced that early allocation of human resources (and time) in the first phases (0 and 1) will yield significant savings in subsequent phases due to fewer design/manufacturing iterations. ·

Product development cost affects a company in two important ways. First, in terms of their specific business strategy, companies must decide the relative significance of new product introduction. The level of effort to introduce a new product will naturally differ from industry to industry. These two considerations, one determined by the company's strategy and the other by the complexity of the products that they design and manufacture, will support general assessments of product development cost. Yet it will be a specific new product concept appearing at a particular time in the company's history that will form the basis for an actual product development cost target.

Such a target has obvious cash flow implications: money needs to be available to be spent. Issues of sources and uses of funds naturally arise, as do their personnel equivalent (numbers of people and skill levels). Other ongoing budget commitments may impinge on the setting of this cost target, despite the fact that the actual cost turns out to depend on the specifics of the proposed project itself. Furthermore, the product development cycle (as defined previously) will probably overlap two or more annual budgetary cycles. This factor tends to enhance the competition for funds, particularly in the current and upcoming budgetary cycle. In other words the timing of budgetary commitments can turn out to be as important a consideration as the total cost of the project. While budgetary realities are hard to eliminate, management should strive to reduce unnecessary and artificial fiscal constraints on new product development. Being obsessed by return on investment (ROI) projections for NPI projects, rather than strategic assessments of value of the proposed new product, hinders the setting of appropriate funding priorities.

A typical financial ambiguity here is whether the budget for the project reflects the full set of costs that will be incurred in its execution. This issue is directly tied to the earlier discussion on when an NPI project really begins and ends. In most companies, the formal project budget will not include all of the up-front efforts to define the initial product concept (including market research and technology assessment). Nor will it include all of the debugging necessary to reach steady state production capability. Perhaps more significant, the budget may not even include all of the time spent by management in planning and reviewing the project and by company troubleshooters called in for informal consulting. To the extent that suppliers are vital contributors to the design and development effort, the formal budget may not fully reflect the costs of their participation.

Given these realities, senior managers must take care to understand the content of any projection of product development cost. Project managers, likewise, need to anticipate what comes automatically with the approval of a project budget. Clearly, it is important that product development cost targets be set and approved. But this is just the beginning of dealing with resource allocation issues associated with the project.

As with the setting of product development cycle times, cost targets can be ambiguous. The question that we have raised about the nature of timing, budgeting, and costing will have different answers in different companies. The answers may even differ across the divisions of a single company or vary through time. The important point for both cycle time and development cost targets, however, is not whether the company has the best definitions, but rather whether those definitions are clearly stated and understood by all. For purposes of continuous improvement (the main topic of Chapter 13), it is also valuable to maintain consistent definitions over time.

Dimensions of Quality

Product quality, as suggested above, is a misleadingly simple term. It is easy to state but can be very difficult to define in particular instances. The notion that there are multiple dimensions of quality is now well understood (and will be discussed later). Setting product quality targets early in the project provides a common vision for all those participating in the design and development activities. At some point in the project, before key design decisions are made, it is desirable to have targets for each of the various dimensions of quality that apply to the product to be developed. Some of these targets will be deemed to be so critical that they ought to be identified and justified even before the formal beginning project (i.e., in Phase 0).

New product quality can be the most ambiguous of the four types of targets. Now that product quality is generally accepted to be a key basis for competition in most industries, defining this term clearly and comprehensively becomes especially important. David Garvin (1987) has defined eight dimensions of quality as follows:

1. **Performance:** Product's primary operating characteristics.
2. **Features:** Supplementary characteristics of a product.
3. **Reliability:** Probability of product failing over time (mean time between failure).
4. **Conformance:** Meeting established specifications.
5. **Durability:** Measure of product life (to replacement).
6. **Serviceability:** Ease-of-repair (downtime, mean time to repair).
7. **Aesthetics:** The look, feel, sound, and so forth, of a product.
8. **Perceived quality:** Subjective reputation of a product, which includes aspects such as ease-of-use and product integrity.

It is much easier to understand this multidimensional definition of quality in the abstract than in any particular situation. The case illustrations in Chapters 7–10 present four company situations where quality was defined in various ways. Other references to the definition of quality in particular instances are made throughout this book. For now, it is most important to appreciate the risk in providing a definition that is overly restricted, unnecessarily ambiguous, or not centered on what the customer would perceive as important.

Ultimately, quality is measured externally in terms of customer satisfaction. For this reason, customer input in defining these internal measures of quality must aggressively be sought. Achieving a high-quality product that has little or no value to potential customers can be avoided by a deliberate and careful customer orientation. Some dimensions of quality are mandated by law and are therefore nonnegotiable. Agencies of government dealing with matters of health, safety, and environmental protection regulate the introduction of new products by setting minimum standards that must be met. As these minimum standards get more demanding (e.g., higher estimated miles per gallon for automobiles), product designers need to factor them in as constraints to be met.

The relevance of all of these dimensions of quality and the priorities among them need to be identified in the concept development phase of a project. Specific measures for these quality targets should also be identified to the extent that this is reasonable. Measurable quality targets of all types can serve as a basis for assessing different design options for both the product and its associated manufacturing, distribution, and aftersales support processes. This can be accomplished either informally or through structured design approaches such as quality function deployment (discussed in the appendix to Chapter 6).

Unit Cost

The unit cost of a product is perhaps the most obvious target to set early in the life of a project. From a marketing viewpoint it makes little sense to set product quality targets without simultaneous reference to price. While traditional industry profit margins will vary, executives generally know the relationship between their pricing strategy and cost structure. Accordingly, product design and development efforts aimed at particular market niches need to be guided by a suitable unit cost constraint.

From the beginning, the core team needs to establish a unit cost target that is compatible with specified measures of quality, associated kinds of manufacturing processes, major product components, and materials. This means that the full set of assumptions regarding the product and its production must be explicitly tied to any unit cost estimate. Being this explicit regarding the unit cost target allows for its reconciliation with product quality options. It also provides an overarching metric for assessing relationships between product design approaches and manufacturing and sourcing options. Decisions regarding a product's materials, and the number, type, and configuration of its parts can be translated into the necessary supplier arrangements and manufacturing system. The level of investment in new manufacturing processes and facilities also has a significant bearing on the unit cost equation. Cost accounting expertise is clearly required at this point and many companies have their own automated cost projection models to produce these critical estimates.

Management must ensure that the relationship between the unit cost target and the volume manufacturing process has been clearly articulated. If this is not specified, the risk of unanticipated costs and delays can be significant. Some companies adopt the convention that "unit cost" targets apply to a point when there has been adequate manufacturing experience in the volume production of the new product. Here, unit cost assumes the existence of some particular "learning curve." An example of this costing convention is given in Chapter 10, with reference to the complex and low-volume manufacture of jet engines. Many electronics products also depend on lowering unit costs through the learning curve.

In contrast, some companies are becoming more aggressive in their new product development strategy and replace learning curve projections with a "mature first-cost" philosophy. This approach is important where markets are competitive, profit margins are low, and the new product needs to get rapid acceptance. Under these conditions, no learning curve is assumed because the product is designed to be easily manufacturable from the start and production processes are thoroughly tested at the pilot stage. In other words, such NPI projects (including the Chapter 8 case from Northern Telecom) intend the unit cost to be "mature" from the start of volume production and therefore remain essentially constant until subsequent design changes are made either to the product or its manufacturing process. Achieving mature first cost requires early agreement on the cost target, tough project management, and careful attention to limiting the use of unfamiliar technologies.

Balancing the Targets

Time, cost, performance, and quality targets should be derived from a careful consideration of how to balance customer, technology, and business requirements with internal capabilities, resources, and priorities. Achieving this balance to achieve a good fit with customer and business needs at both the product line and higher levels is a central strategic challenge.

The four types of NPI targets are mutually interdependent and must be balanced with each other (see Figure 3–2). As one modifies any one of these targets in a new direction, one has to think of the repercussions it may have on the other targets for that NPI project. A key, though often underestimated, factor in managing product design and development projects, is the ability to set targets that are in equilibrium relative to each

FIGURE 3–2
Balancing the Target Set

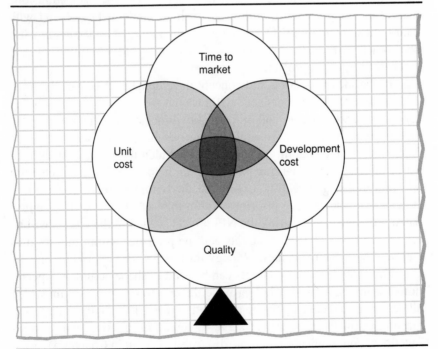

other. Overall customer satisfaction, or even delight, is an overarching goal to be constantly kept in mind when seeking the optimum balance among the individual targets.

All targets should be considered simultaneously, and the resulting target set should be an integrated reflection of the relative priorities among targets. Trade-offs among the several targets must be explicitly considered. Accordingly, it is a mistake to have different people set the various targets. Nor should the targets be established in any fixed sequential pattern, arbitrarily or by edict (e.g., "Get 30 percent of the costs out." "Get the time down by 40 percent").

Target setting for NPI projects is more than a planning exercise. It is the fundamental basis for guiding product design and development activities and for exercising the project review and control functions described in Chapter 2. Accordingly, it is essential that all involved personnel understand the general implications of setting each of these kinds of targets. Each of the four types of targets has its own unique considerations, some of which have been hinted at previously. To appreciate their strategic significance and the richness of issues embedded in their selection, one must deal more explicitly with each of these kinds of targets and with their typical interactions. Once this conceptual level of understanding has been reached, one can turn to issues inherent in setting particular target levels for a specific project.

In the idea validation and concept development phases of the project, the core team members need to thrash out the major trade-offs among time, cost, performance, and other quality dimensions. After considering factors of technical risk, manufacturing capability, and other resource constraints and availability, the team should identify a single, integrated set of targets. Benchmarking exercises can help in defining targets with adequate reach and reality. Central to this entire effort is the explicit statement of the relative priorities within this set of targets: If unanticipated problems occur during the NPI project, which of the targets will be reassessed first? The set of targets and the priorities within this set thus embody the existing business strategy for this part of the product line.

Taken as a set, these targets can be used to shape common expectations for the NPI effort. They provide the constraint space within which all designers and developers are expected to work. Having these targets in place early in the project can serve to force constructive, early cross-functional communication (a topic discussed in detail in Chapter 4). Most directly, this set of targets can be used during the course of the project as the primary basis for assessing design trade-offs (see Chapter 5). These targets can also be used as a basis for the retrospective assessment of the product design and development process, a critical step toward continuous improvement. Given all of these considerations, it should not be surprising to find that leading product innovators routinely formulate such a set of targets.

Regardless of their size and position in their industry, companies wishing to impose more discipline on their NPI process should start by thinking in terms of such targets. Perhaps a well-endowed start-up company can afford not to measure product development cost very precisely, but even then some rough guide along this line will ensure that there is no misunderstanding about the order of magnitude of that constraint. Similarly, a market leader may have some slack in terms of the time when the next product innovation must become commercially available. Even when speed to market is less critical, time, cost, and quality targets are needed as a basis for developing a realistic project schedule. To ignore such targets is to accept product design and development as a random event. No company can afford such a "hands-off" attitude and an undisciplined process. Furthermore, management must encourage these targets to be set in conjunction with a detailed scope of work. It is a mistake to promise customers that one or more of these targets will be achieved until the product concept and associated work plan have been developed.

SETTING THE TARGET LEVELS UNDER DIFFERENT SITUATIONS

Target setting for new products is a social act of responding to perceived pressures. The larger the company, the more complex that act will tend to be. While it is easy to find documentation on what the targets were, it is often less clear as to how they came to be. Without getting into a clinical level of detail, this section discusses variations in the patterns of setting particular levels for the various types of targets (i.e., shortened).

The target for development cycle time is driven down by concerns about possible competitors' preemptive moves, coupled with estimates of increased revenues and profits that accompany every additional month that the planned new product arrives "early" to market. Quality targets, in any of the eight dimensions, are driven up to meet images or forecasts of heightened customer requirements. Competitive considerations blend with perceptions of market demand in forcing certain unit cost targets. All three of these targets may set the stage for setting the fourth target, the product development cost, in light of perceptions of the fiscal health of the enterprise and the relative significance of other potential uses for the same funds.

Development cycle time targets are normally suggested by the marketing arm of a company when there is perceived urgency in introducing a new product. Such a sense of urgency is now widespread and will continue to increase. As mentioned in Chapter 1, time-based competition is becoming increasingly common across many industries, and this trend is not likely to disappear. Indeed, it is quite probable that the notion of competing on time has already become part of the culture of most companies with products that have a high-technology content (similar to the orientation toward quality before that).

The priority of a given NPI project and the perceived cost of being "late" to market will directly influence the cycle time target (see Figure 3–3). The basic economic justification for this line of thinking is that much of the potential sales revenues (and profit) from a product with a short life cycle will occur during the lead time that a company establishes between its own launch date and that of the competition that follows. Delays in product launch cut into that key revenue potential to the point that a late launch can be responsible for the economic failure of a product that could have been a winner.

Looking more closely at the setting of such time targets, one finds a number of forces at work. As shown in Figure 3–4, some forces are external to the company: perceived market requirements; a request from a major customer ("lead user"); the projected availability of a new key technology; announced (or anticipated) enhancements or new product releases by competitors; a new regulatory mandate; the scheduling of an annual trade show; or even pressure from the financial community. Other forces that affect speed to market come from within the company or business unit and include product strategy, mandate from a senior executive, or other concurrent priorities.

The notion of setting a development cycle time target may seem intuitively appealing but this is not always the case. In a "technology driven" company, for example, design engineers may be inclined to let the nature of the problem guide the time they spend to reach an acceptable

FIGURE 3–3
The Cost of Missing the Target Introduction Date

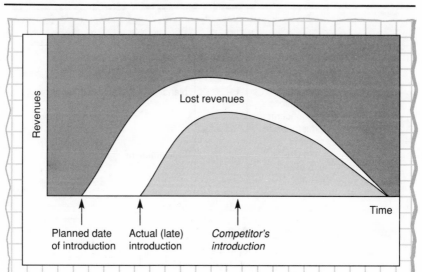

solution. Such experts might tend to resist time targets set by others who claim to reflect the needs of the marketplace. Setting a single development cycle time that will be accepted by all participants in the project as a target that they are capable of achieving may prove to be a challenge in itself.

The development cycle time (often called the *lead time*) target shapes the context within which all project management takes place. The target is usually stated in terms of a "due date" for the commercial availability of the new product. The logic for setting the due date will vary, as will its significance. In certain industries and competitive contexts it will be clear from the start that a particular date for commercial availability of a new product needs to be established as a top priority target. In other situations the due date is a target to aim at, but it is generally understood that achieving this target would be nice but not necessary. A few of the more typical situations are as follows.

FIGURE 3–4
Factors Affecting Cycle Time Targets

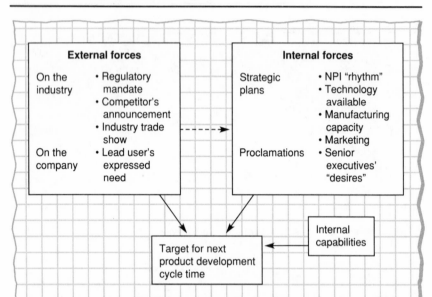

The Urgent and Inflexible Due Date

It is not uncommon for one company's product to be a component of a larger product which already has a firm date for commercial availability. Here, the viability of the new product introduction is conditioned on being accepted as a supplier to the downstream customer.

Aircraft engine manufacture is a classic example of this situation. Airlines commit to the purchase of an aircraft of which the engine is a dominant determinant of performance. An airframe manufacturer (e.g., Boeing), accordingly, must select an engine manufacturer far enough in advance for the engine to be commercially available for use in the new aircraft itself. If an engine manufacturer does not have a suitable engine already available, the time available for new product development is clear.

Such was the case (described in Chapter 10) of GE's CF6-80A commercial jet engine, where the key to market entry was to schedule the development effort to be compatible with the introduction of the Boeing

767 plane. When executives of the GE Aircraft Engine Division made their go-ahead decision, the required NPI cycle was four years. Since a new aircraft engine normally took seven years, this ambitious time-to-market goal drove many subsequent design, development, and manufacturing decisions.

The Technology Leader without a Commanding Market Share

Another time-critical setting is when a company's strategy is to stay at the technological forefront of their industry when they do not own a commanding share of the market. At Amdahl Corp., a major producer of high-performance computer mainframes, for example, a product research group will forecast when competitors' advanced products are likely to be introduced, thereby identifying the cycle time target for their next mainframe computer. Such a company has so much at stake in every new product introduction that they cannot afford to miss a projected market window.

The Technology Leader with Major Market Share

In contrast with the above example, consider the situation when time to market is not as critical to the success of an new product. For example, suppose a company is producing a line of products for a particular market and already has a strong position in that market. A new project in this arena may have a time-to-market target based on considerations such as manufacturing capacity, technological readiness, competitive developments, or resources available. It is especially common for companies to allocate more lead time when priorities and resource availability indicate that the new product effort cannot receive the full budget over a shorter time.

One example of this was Motorola's development of the Keynote pocket pager (a full case study is presented in Chapter 9). Motorola already had a significant market share and there was no impending competitive threat. The company wanted to cancel some of its older, less cost-effective products. Accordingly, when it initiated the project to design and develop Keynote, it was commonly understood that the cycle time target was less of a priority than meeting their extremely high and unyielding quality standards. It was not considered to be a major failure when Keynote was introduced considerably beyond the planned release date.

The Start-up with a Narrow Window of Opportunity

A new company in a highly competitive and dynamic industry typically faces a narrow window of opportunity. Either they introduce a product with significant market appeal and profit possibilities in their first attempt, or many will collapse from financial pressure. Since such a company, by definition, lacks a loyal customer base, they need to take market share from existing companies in the industry. This means that their initial product has to be better than the competition and well-timed to reach the market before a significant competitor offers a comparable (or even better) product of their own. Defining customer requirements and competitive behavior in an unfamiliar industry is, by itself, a difficult act of planning. Establishing an NPI effort and executing it on schedule may be even more difficult. Such was the situation facing the NeXT Corp. (see Chapter 7), when it was founded by Steven Jobs to build a radically advanced personal computer. (NeXT, however, was a somewhat special start-up in that its considerable source of start-up funds allowed it to continue to work on new products even after its initial offering failed to become a winner in the marketplace.)

The Turnaround Situation

A company facing the impending loss of a market niche will sometimes try to set an overly aggressive target for introducing a new product. Early assessment of the development requirements for this new product will sometimes lead management to revise its original target introduction time, rather than to risk introducing a product with limited market potential.

Such was the case of the Norstar product at Northern Telecom, Inc. (see Chapter 8). Marketing had originally determined a target introduction date for Norstar based on the need to offset declining sales in its existing Vantage product, but accepted a target date one year later, as soon as the realities of the new project and the expected time of the competition's new offerings became available. As an interim measure, Vantage was enhanced to extend its market for a prespecified period until Norstar would become available. The implication of this example is that the setting of a development cycle time target can be a deliberate strategic act; as more information is gained on flexibility in market time windows and on variability in the time required for technology development, one can make a more informed assessment of a "reasonable" cycle time target. Such an approach to the reduction of risk, of course, has its own costs, including the time to conduct such an iterative analysis.

INTERACTIONS AMONG THE TARGETS

Balancing the targets can become a challenging part of Phases 0 and 1. In particular, anticipating the connections between cycle time and the other targets deserves special attention. We illustrate this challenge by identifying some of the more significant trade-offs and how the extent of simultaneous engineering can be important in these types of trade-off considerations.

Cycle Time and Development Cost

Two kinds of trade-offs between these two variables have been observed. First, the shorter the project, the less expensive it is likely to be simply because there is less time to spend development funds. In other words, tight scheduling puts caps on the number of hours that can be spent on various planning, design, and review activities. Development costs can likewise be bounded, although the extensive use of overtime and expensive vendors are common responses to tight time schedules. The fact that shorter projects and higher productivity seem to go together in Japanese auto companies was documented by Kim Clark and Takahiro Fujimoto (1991, chap. 4).

Second, viewed from another perspective, higher development costs may lead to longer cycle times under conditions of resource scarcity. This situation can easily arise in a cyclical industry (e.g., machine tools) in the face of a downturn in sales. A similar situation of scarcity can arise in companies that are anticipating growth and have already allocated almost all of their available funds to various forms of innovation and expansion. In these and other instances, getting the funds for a large product development effort may require shifting some of the expenditures to a later budget year than would otherwise have been necessary.

Cycle Time and Product Quality

As is true in the operation of motor vehicles, speed limits ought to be observed in the design and development of products also. If management becomes obsessed with speed to market, the cycle time target for a project may be set unrealistically short, to the point that design errors occur, and development and testing corners are recklessly cut. In Chapter 2, an example of this sort was described, involving a compressor for a new model of refrigerator. This example and others like it suggest that blatant violations of such speed limits invite problems of quality. Which dimensions of quality are affected will depend on the problems that arise so it is hard to know in advance the types of quality targets that are most endangered. To make matters worse, it is not usually obvious in advance what the true speed limits will be.

This line of thinking suggests that managers should try to avoid setting cycle time targets and product quality targets both of which simultaneously push the boundaries of proven ability of the organization. Unfortunately, the natural corollary to this proposition—that generous cycle time targets should facilitate the achievement of ambitious product quality targets—does not necessarily hold. There is simply no guarantee that the extra time will be used to achieve better product quality; it may, for example, lead instead either to a more leisurely paced project or encourage unproductive bureaucratic struggles among the groups involved. In general, an inadequate amount of resources seem to be allocated to the concept development phase. Adding to this component of cycle time, and using this time for constructive product planning and broad problem solving, ought to increase the probability that the associated quality (and cost) targets can be achieved.

Cycle Time and Unit Cost

As with other trade-offs already mentioned, the primary relationship between the cycle time and unit cost targets is through the design activity itself. To the extent that a longer planned cycle time facilitates more early involvement of manufacturing considerations in the early stages of product design, it becomes more likely that the unit cost targets for the new product will be met. Early manufacturing involvement should promote design-for-manufacturing techniques, which (see Chapter 6) can lead to fewer parts, easier assembly, less scrap, higher yields, and other factors that combine to lower the unit cost of a new product. Accordingly, companies committed to the objective of a mature first cost must be willing to spend more time and resources in the early phases of the process where critical design decisions are made.

This trade-off of development time for reduced unit cost is also true to some extent for the portion of cycle time devoted to prototype development and testing. Here time can be well spent in the search for potential manufacturing problems and in design changes in the pursuit of lower unit cost and, perhaps, higher quality.

These disciplined design and development practices may lead to the best of all situations, where both cycle time and unit cost are both reduced simultaneously. After all, product and process redesigns (often called *retrofitting*) in the pursuit of cost reductions is not a law of nature. Getting the design right initially may save time as well as promote the achievement of unit cost targets and customer satisfaction. This is, after all, the dual objective of enhanced planned use of simultaneous engineering.

Building Simultaneous Engineering into NPI Schedules

Cycle time targets for NPI will often determine the amount of simultaneous activity that is necessary. By adopting "critical path" types of analysis, one can easily see how firm and urgent cycle time targets can tend to predetermine the structure of an NPI effort. The pursuit of such "simultaneous engineering" approaches to product design and development (see Chapter 2) is now widespread. While simultaneous engineering offers clear benefits up to some point, in terms of reduced cost and improved quality, there are limits on the overall benefits to be gained from arbitrarily shrinking the cycle time target. Furthermore, there are organizational costs associated with the initial implementation and subsequent enhancement of simultaneous engineering approaches. In the extreme, one can conceive of the situation where demanding too much simultaneity can become counterproductive.

Avoiding Overly Optimistic Assumptions

It is easier to say that NPI time targets should be set appropriately than to do so. The net effect on the product development cycle time of a set of forces external and internal to the company is hard to project with much accuracy. In addition, ambitious ("stretch") time targets may be needed to force the product development team to overcome common unproductive practices and procedures. Nevertheless, managers should strive to avoid setting unachievable time targets. Two of these traps, involving overly optimistic assumptions, are related to resource availability and a common set of project complexities and uncertainties.

There is a natural tendency to overlook the realities of resource and technology availability. This includes the identification, selection, and possibly relocation of appropriate project personnel. It also typically includes delays caused by lengthy approval processes associated with these activities or with the budgetary decisions preceding them (although companies should try to remove such delays). In addition, in companies where NPI and other projects do not end on schedule, a new project involving the same personnel is apt to start later than originally intended. Some of these delaying factors will be apparent to experienced company personnel who wish to develop a realistic commercialization date for a new product. Getting a late start, based on naïve assumptions, is a poor way to begin a time-managed project.

There is a danger of not factoring in special project complexities and uncertainties. Some projects are inherently more complex and less controllable than others as, for example, when the responsibilities for design are split among different groups, located in different places, with little prior joint working experience. This is also true whenever the key resources on an NPI team are being shared with other projects. Industries subject to governmental regulation are exposed to an additional dimension of complexity and uncertainty in being subject to mandated reviews or approvals in the middle of a project. In the United States, this occurs when special issues of public health or safety accompany the product, as in the role of the Food and Drug Administration (FDA) in regulating pharmaceuticals or the Federal Aviation Administration (FAA) in the field of air transportation. Because such reviews are conducted by external government agencies, their timing and outcomes are harder to anticipate than internal company "gate" reviews.

Some companies have a strategy of being a "close follower" of the industry leader. Soon after the leader introduces a new product, the follower will offer a more cost-effective version of that product. Such companies may experience unexpected delays in designing their own new product while they await a product announcement by the traditional leader. Finally, some projects will face unexpected delays due to technology problems that were either not anticipated or greater than expected without any contingencies having been provided. What may have been originally planned as straightforward development engineering may turn out to be new terrain with time-consuming, problem-solving activities.

Many of these factors will be apparent at the beginning of a project to anyone who directly considers them. It is better to take an educated guess on their impact on the development cycle time than to leave these factors out of the equation. To ignore them is equivalent to betting that they will not occur.

Establishing Priorities among the Targets

A cornerstone of project planning for product design and development is specifying the priorities among the various targets. The questions that need to be answered are straightforward but may lead to considerable debate among project planners, managers, and team participants. What target will be relaxed first when things start to slip beyond the point of full recovery? Quality is rapidly being considered a nonnegotiable target in many product markets. But what are the various measures of quality and what are the ranges of acceptance? How much slack is in the cycle time target? How much is the company willing to rely on a cost reduction effort after a new product is released?

Questions such as these should be asked and answered no later than in Phase 1. The priority ranking of the time, cost (or price), and quality targets must be explicit and known by all who participate in the project and those who serve on the executive review committee. As you will see in the case histories presented in Chapters 7–10, each had its own distinct set of priorities, which guided certain decisions during the execution of those projects. Whenever possible, companies should use potential customers to help assess trade-offs among the various targets and set priorities among them.

MANAGING NPI PROJECTS TO MEET CYCLE TIME TARGETS

Regardless of the specific product and manufacturing process being developed, the project manager and the executive review committee should pay attention to the progress of product design and development. As described in Chapter 2, this is often done in the context of formal phase reviews. Even though the project manager monitors progress on a contin-uous basis, individual project participants need to be sensitive to the time-related issues of their ongoing work.

Executives participating in formal phase reviews of the NPI effort must also integrate considerations of timeliness with assessments of the decisions and progress made to date. Sometimes time itself will be the primary focus of attention: How late are we? What can we do to catch up? In other situations, as new issues and options arise, managers must assess the associated time implications along with other factors.

The most critical aspect of cycle time management is the reduction of variability in time, cost, quality, or performance arising from risky and unpredictable factors (i.e., new technologies, tooling development, ven-dor start-up, software development, new organizational structures, and human resource recruitment or development). While it is difficult to pre-dict exactly which potential bottlenecks will materialize, it is possible to plan so that their materialization does not disrupt the entire program. Major planning tools include:

- Contingency plans to back up critical path items having high-per-ceived variance.
- Off-line advanced technology development in both product and process technologies that produces relatively mature technologies that will be ready to feed several new projects when needed.
- Off-line infrastructural development, such as implementation of new design technologies and training in their use.

The ability to manage NPI projects to meet such cycle time targets will depend on a set of structural "enablers" and "barriers." Enablers are conditions that promote time-based management in the context of a new product introduction. Barriers have the opposite effect.

Three Enablers for Meeting Time Targets

Regardless of the specific requirements of a project, three types of capa-bility will enable the achievement of a target for its development cycle time:

1. Company culture embracing time-based management.
2. Allocation of people to the project needed.
3. Back-up options to overcome projected problems.

A Company Culture that Embraces Time-Based Management.   The importance of cultural factors to the overall success of an NPI project will be discussed in detail in Chapter 4. Here we list just those aspects most closely tied to meeting planned schedules. In some industries, developing new products to meet customers' procurement schedules is so critical that seasoned employees of successful companies naturally accept this un-yielding priority. Managers in other companies may find meeting pre-established time targets to be an uphill struggle; everyone on the NPI project seems to come up with a reason to extend the original schedule to allow more time to solve some particular problem. Here, managers must take special steps to get key NPI participants to accept the stated time targets, including the basic design and development schedule. Over time, the tools that these companies use to get formal buy-in shape a culture where the deadlines are taken seriously by all who are involved. Time-based management is a philosophy which, if aggressively pursued in all areas of company business, will tend over time to create an important cultural enabler for effective product design and development.

The Allocation of People to the Project as Needed.   The most "urgent" product development project is not always the activity with the highest priority and, therefore, may not be granted first call on human resources. Under such conditions, there is likely to be a gap between the required and actual levels of experience, mix of skill, and number and timing of those assigned to the project. This has direct negative impacts on the ability to meet preset time targets. However, in situations where such a priority does exist, the achievement of target deadlines is significantly enabled. One such situation is where the company is betting its future on the upcoming new product. This is always true for a start-up company such as the NeXT Corp. depicted in Chapter 7. In addition, it is often true for a company with a small number of products and rather infrequent and large new product introductions (see GE Aircraft Engines, Chapter 10).

Back-up Options to Overcome Projected Problems.   In some industries, the procurement or development of tooling for manufacturing a new product or component is invariably on the critical path of the NPI project. When the project is under severe time pressure, the temporary use of soft tooling (eventually to be replaced by hard tooling) can facilitate speed to market. Tooling made of softer metal does not last as long in use as that made of a harder substance, but it is usually easier to develop, procure, and modify in a short amount of time. The use of soft tooling for developing prototypes and doing pilot runs often justifies this extra procurement effort and cost. Another example of flexibility is to design a new product using existing parts in stock rather than going through a lengthy procurement cycle for a similar but new part. Later in the NPI process, establishing dual sources for critical components helps ensure adequate on-time deliveries of materials.

There are also options to increase the extent of planned simultaneity in the product design and development process. This topic was discussed in Chapter 2. One special option in this regard is for a company to release a new product design for production before it is fully certified by a regulatory agency; clearly, in such a situation one must assess the risk that certification will be denied.

Five Barriers to Meeting Time Targets

Achieving the desired development cycle time is, of course, never guaranteed. Five typical barriers to meeting such time targets deserve special consideration by managers seeking to improve the resulting time to market.

1. Excessive time pressure.
2. Inadequate buy-in on time, cost, or quality targets.
3. Inadequate consideration of exogenous factors.
4. Loaning key people from an ongoing design and development project.
5. The self-fulfilling prophecy.

Excessive Time Pressure.   The imposition of time pressure, in the absence of prior successful experience, may generate a climate where participants become defensive and parochial rather than risk-taking and collaborative. In such situations one often sees more professional attention being paid to justifying decisions (leaving a bureaucratic paper trail) than in working on better solutions. Such behavior easily turns out to be self-defeating. It suppresses the early identification of problems. When these problems are identified later in the NPI effort, they invariably take more time (and cost) to remedy. This is less apt to occur in a company that traditionally works under time pressure to produce products where performance objectives force collaboration across functions and specialties. Chapter 4 discusses this issue further.

Inadequate Buy-in on Time, Cost, or Quality Targets.   In the interest of time, some companies intentionally limit the early involvement of those who would be impacted by the initial assumptions regarding unit cost or manufacturing process investments. This practice can easily backfire and lead to more, rather than less, project delays. At later stages of product design and development, unanticipated demands for financial support, such as "management reserve" accounts, supplemental requests for investments in special manufacturing processes, often need to be addressed. At this later point there is always a risk of critical time delays while resource controversies are resolved.

Anticipation of similar time delays can lead a company to cutting short either the degree of customer or supplier involvement in early phases of the project or the subsequent extent of product testing. The philosophy here is to get the product out the door and to fix it later. In the drive to keep the project "on schedule," team members may not recognize the risks that are being taken. A more realistic cycle time estimate can avoid such undesirable situations.

Technological risks are more likely to be reasonably perceived when the most experienced technical experts are assigned with no constraints on getting additional technical inputs as needed. In this situation there is a different danger: that excessive early management pressure to meet the product introduction schedule will result in less technological risk-taking and creativity.

Inadequate Consideration of Exogenous Factors.   Factors external to the company may be more significant than internal factors in raising barriers to the on-time introduction of a new product. A common source of such problems is the supply of a key component or material. Whether it be a new microprocessor chip with inadequate industry capacity or a rare metal in scarce supply, the development plans of end products can be unavoidably delayed. In the fast-moving electronics industry, products get designed based on components that are still in the R&D stage in the supplier organizations. Here, the end product's estimated time to market can be delayed simply due to the unanticipated problems in the development of the requisite components. Other exogenous factors that need to be explicitly projected and monitored include new competitive offerings and technological breakthroughs, both of which may change customer expectations.

Loaning Key People from an Ongoing Design and Development Project. Whenever this course of action occurs, it seem to be the lesser of two evils, because the employees' contribution on the new assignment (e.g., redesign of a module in the field, dealing with unanticipated manufacturing problems) is deemed to be even more critical than whatever they were doing on their original NPI project. This barrier can only be reduced through more effective human resource planning and scheduling. Nevertheless, a natural tension seems to exist, across industries, between the priority of new product development and the servicing of existing products.

The Self-Fulfilling Prophecy. Finally, there is the high, long-term cost of habitual overreliance on unrealistic NPI cycle time targets. When such targets are never met, no one is inclined to pause during a project to identify lessons for the future improvement of the NPI process itself. Furthermore, repeated experience working under unachievable time deadlines leads participants to discount at the outset the significance of such deadlines. This creates a self-fulfilling prophecy where the deadlines will have to be extended even further into the future.

Assessing the Feasibility of Cycle Time Targets

The feasibility of achieving development cycle time targets directly depends on four types of factors, each of which may be assessed in advance: (1) the extent of simultaneous engineering; (2) the quality and familiarity of the NPI infrastructure; (3) the likely claims that will be made on key resources; and (4) the inherent risk and uncertainty.

Simultaneous Engineering. If there is not enough simultaneity in the design of an NPI project, the risk of the "over-the-wall" syndrome (see Chapter 4) leads to errors and delays due to recycling to earlier phases of the NPI project. The opposite danger—that the project is forced to engage in more simultaneity than is customary—is of growing concern as time-based management rapidly becomes part of the conventional managerial wisdom. Then, as more and more of the key product and process development activities are done in parallel, rather than in a logical sequence, the risk of error begins to rise. In either case, errors take valuable time to correct, leading to the possibility that corners are cut during the latter stages in a valiant attempt to catch up to the original target for commercial shipment of the new product. When activities involving product testing or process validation are rushed, the risk can turn out to be quite unacceptable.

NPI Infrastructure.   This factor can make a big difference in pursuing an aggressive cycle time target. Design technology such as CAD/CAM and CIM (computer-aided design/computer-aided manufacturing and computer-integrated manufacturing, respectively) can support extremely rapid transitions from design and development to volume production. Yet this is only liable to occur when the technology is well-established in use at the company. Ironically, such technology is often introduced as part of a new product development effort, thereby generating additional potential complications and delays for the project.

Supplier management capabilities are also part of the NPI infrastructure. Strategic use of suppliers as part of the design and development effort can bring an otherwise impossible time target within reach. Suppliers who are part of the concept development process can offer valuable design suggestions early, thereby avoiding wasted effort. They can start their own process development activity earlier thereby providing needed materials, components, or tooling sooner in the prototype testing and pilot production phases. Unfortunately, the annals of new product introductions are filled with accounts of how a key supplier was responsible for making the whole project late.

The key role of purchasing in the introduction of new products also deserves special mention. Regardless of whether the target of greatest concern is time to market, quality, or cost, purchasing can affect the outcome. With respect to lead time reductions, getting purchasing involved early in the qualification of vendors can save valuable time. Purchasing ought to have people with the experience and skill to select vendors that can meet the requirements of a new product (or part of its production process). If the NPI project does not have a purchasing expert on the core team, then getting the services of such a person as needed should be a high priority.

As described in Chapter 2, a well-honed executive review process also supports the achievement of a cycle time target. However, when such a process is not well-executed it too is likely to create an additional layer of delays and unnecessary barriers to speedy launching of a new product.

These several examples suggest that there are two relevant dimensions of NPI infrastructure: administrative and technological. The administrative side is discussed more fully in Chapters 2 and 4. Issues in the management of technology are dealt with in Chapters 5 and 6.

Claims on Key Resources.   A top priority NPI project in a well-endowed company has a lot going for it. The project priority will dictate whether the resources originally allocated to the new product introduction are apt to be sent to other unplanned assignments during the life of this project. Even when the product and production process are complex, a high-priority project, perhaps under a bet-the-company situation, can be executed in record-breaking time.

Risk and Uncertainty.   Companies working on a new product in one of their traditional markets, and based on the technologies in which they already have proven expertise, will naturally expect far fewer surprises and delays than when this is not the case. Familiarity with the market and the key technologies leads naturally to design choices that probably can be implemented within the available time limits. Unfortunately, the dynamics of today's markets and technological progress is such that this degree of familiarity may be less common than has been traditional in many industries. Risks from underestimating the difficulty of solving identified problems and uncertainty from problems never before encountered will arise. Whether such occurrences will render cycle time targets infeasible depends on their extent and severity and the response capability built into the NPI plan.

THE ROLE OF THE PROJECT MANAGER

Now that we have reviewed the structural dimensions of project management, it is time to turn to the personnel side of this subject. Here we deal with the role of the project manager. In the next chapter we concentrate on the project team.

Project management is not a new activity, nor is it restricted to the field of new product design and development. In fact there is considerable literature already available on this subject (see, for example, Kerzner, 1989), although very little of it speaks directly to the issues most central to successful product design and development. Rather than trying to summarize everything about this subject in the limited space available here, we identify and discuss two matters of direct interest: responsibilities and skills needed.

Responsibilities

The project manager, in many companies, is charged with providing leadership in meeting the targets. When this leadership is exercised through the full range of program phases, it is liable to include responsibility for:

1. Defining the product requirements, and the costs and profitability associated with its introduction.
2. Determining the viability of the product concept from the company's point of view.
3. Working with functional managers to select a project team.
4. Controlling the project's budget.
5. Managing discussions and negotiations involving trade-offs in project goals and targets.
6. Being a champion for the new product and a role model for team members.
7. Leading the team to accomplish the goals of the project.

Whether a project manager is given this full set of responsibilities depends on the company's structure and management strategy. Clark and Fujimoto (1991, Chap. 9) refer to "heavyweight" and "lightweight" project managers, where the former is a general manager of the product being responsible for the project outcomes and, possibly, the direct supervision of a set of dedicated project resources, and the latter is more of a facilitator of progress and an indirect coordinator of resources that remain assigned to their traditional functional organizations.

Skills Needed

It is desirable that a heavyweight NPI project manager have a business focus and considerable experience, especially in the functional areas most critical to the project's success. Personal skills desirable for such a project manager include decisiveness and sensitivity to meeting deadlines, balanced by the ability to coordinate and monitor diverse activities, and motivate teams to do collaborative problem solving. The project manager is also typically charged with developing the technical and managerial skills of team members.

Sometimes it is difficult to find one person who can successfully discharge all of these responsibilities. In such cases, senior management must be aware of the demands on the project manager in different stages of the NPI process and be prepared to make beneficial changes in project leadership at the appropriate time. For example, different skills are required to lead an engineering team through the conceptual development stage, compared to the subsequent technology implementation stage. In the earlier stage there is a premium on motivating the engineering team to creatively identify and assess various product design approaches. In the later state, the ability to promote closure on design and development tasks within time and budget constraints becomes paramount. In this kind of situation, a company may have to switch project managers midstream if the one who leads the team through the concept development stage cannot make the needed shift in style and substance. At all stages, task coordination is an important skill of the project manager, although the demand for such skills is likely to increase as the project moves into detailed design and prototype development. The product manager should also be skilled in empowering team members to seek innovative solutions within the framework of the common set of targets.

 Effective NPI project managers will also be skilled in communication and negotiation. Protecting key team members from conflicting demands on their time or concerns about their future role in the company are clearly important. Being able to convince senior management to provide continuing resources and support for the NPI is also required, as is the ability to be clear and concise in reporting progress and potential problems. Clear understanding of the business plan for the new product will facilitate such external communication and negotiation.

SUMMARY

Setting targets for the development time, development, cost, quality, and unit cost is an essential part of planning an NPI project. Issues in setting these targets include relationships among the targets themselves and the need to maintain a customer orientation in defining the critical success factors. These issues need to be explored and resolved by the project team. Management at the project and executive levels needs to pay careful attention to this target-setting activity to ensure that assumptions and choices are well-validated, recognizing the limitations of time and available information. The common challenge is to seek evidence that the project targets have captured basic expectations of potential customers, rather than simply summarizing unsubstantiated employee perceptions.

Once the targets are set, they must be communicated clearly, consistently, and repeatedly to members of the extended (internal and external) NPI team. Suppliers, for example, must understand precisely what is meant by *quality* and *timely delivery* (and some of the suppliers should have participated in discussions on these targets). Priorities among the targets must also be made explicit as a guide to subsequent design and development activity.

Special effort is needed to increase the likelihood that such priorities are shared because this is not naturally the case. Different individuals and groups carry into the project differing perceptions about the urgency of time, the reality of deadlines, and the relative importance of the various targets. One of the responsibilities of the NPI project manager is to promote a common understanding of the project priorities among the team members.

The approved set of targets should be firmly adhered to, unless there is good reason to revise it. Sometimes it is appropriate to reassess and possibly revise the targets, as when, for example, information is gained on significant changes in market windows, customer needs or desires, the availability of technologies, or time required for technology development. By instituting mechanisms for review and revision, management can help ensure that such changes are deliberate, and are communicated as such, rather than simply resulting from drift, slippage, or indecisiveness.

In summary, while Chapter 2 described the formal NPI process, this chapter moved closer to the action of product design and development. We looked at key aspects of planning for the new product and its associated NPI process, as well as the ongoing management of that process. Now we are ready to look at the substance of product design and development: the people who do it, and the technologies with which they work. Chapter 4 deals with the former and Chapters 5 and 6 with the latter.

CHAPTER 4

DEVELOPING COLLABORATIVE HIGH-PERFORMANCE TEAMS*

Thus far we have concentrated on the formal side of product design and development, that is, phases of work, reviewing progress at logical conclusions of those phases, and the planning and management of the projects themselves. Ultimately, however, product design and development is a group activity in which people interact with each other on a daily basis while exploring and assessing options, solving problems, making decisions, taking actions, and seeking assistance. The types of problem-solving activities vary from establishing specifications for the product and manufacturing process to the creation, assessment, and validation of design alternatives. We now turn to the more human side of product design and development and to the workplace where experts of various disciplines come together in a team setting.

Three questions need to be addressed:

1. What is important about the workplace culture of NPI teams?
2. What kinds of skills and orientations do team members bring to the activity of product design and development?
3. What can be done to promote the development of collaborative behavior by those on NPI teams?

The importance of a workplace culture that supports high-performance, new product introduction is now generally recognized. Manufacturing organizations of all types are pursuing this goal and progress is being made. Nevertheless, for many well-established and highly regarded companies in the United States, success in this regard has been elusive.

The difference between a cross-functional team that works well and one that doesn't most often lies in the extent to which members have common objectives and are able to create new, shared understandings and

*This chapter draws on material from two earlier publications by the author (Rosenthal, 1990a and 1990b).

meanings. Team members come from different organizational subcultures that may be based on functions, divisions, or geography. Members of each subculture bring their own assumptions, meanings, priorities, ways of thinking, and styles of communicating to the team, and usually assume that their language, styles, and meanings are shared. When these undiscovered differences are not worked through, people leave their meetings and conversations assuming they have been understood. When they discover otherwise, confusion and anger often result, much time may have been lost, and other damage may occur. This chapter reviews findings about workplace culture related to product design and development, describes some of the subcultures that are involved, identifies issues that most commonly arise in this domain, and points to directions for improvement.

A TALE OF THREE COMPANIES

Workplace conditions and requirements may differ from company to company. Consider the following cases of three companies that, in the late 1980s and early 1990s, attempted to introduce a new product under extreme competitive pressure.

> **Company A** was formed to develop a product that would compete in a rapidly growing and highly competitive market for electronic equipment. Its founder selected the key engineering and manufacturing managers with a careful eye to their willingness and ability to work together. In their prior jobs, these managers had experienced the problems of trying to introduce new products when independent functional orientations were more dominant. They joined Company A not only for its significant commercial potential but also for its special professional climate. From the beginning, Company A's corporate culture has emphasized teamwork and integration of work activities in the design of the new product, the selection of suppliers, and the development of a manufacturing capability. The founder of Company A believes that "business has gotten so complex, dangerous and quick moving that no matter how brilliant people are at the top, it's not good enough any more. You have to utilize every brain in the whole organization." Company A seems to thrive on direct constructive communication. Everyone speaks his or her mind

in group problem-solving meetings. Everyone is committed to the design, development and delivery of a top-quality product. To a large extent the product and the manufacturing process to produce that product were developed simultaneously. From the outset, manufacturing involvement in product design decisions was the rule, not the exception. The founder of Company A strongly believed in the importance of manufacturing as a competitive weapon. This company hopes never to have problems with its design-manufacturing interface, although they recognize that if their plans for rapid growth succeed, they will have to be diligent to avoid the onset of rigid and conflicting functional cultures.

Company B had a well-established business in the telecommunications industry when it faced a special challenge calling for radical change in their process of new product introduction. An early, narrow market window appeared when one of the major competitors introduced a product that lacked significant advances and the other most aggressive competitor announced a delay in its next product offering. Company B set out to develop a piece of end-user equipment in a business where they had little previous success; their reputation for quality products was based largely on more complex integrated electronic systems. Either a winning new product had to be developed and introduced quickly or an existing plant that was extremely underutilized would have to be shut down. Facing strong time-based competition, they were able to design, develop, and manufacture this new product with no problems of quality in what was their record time of little more than one year. This product soon became a market leader for its functionality, flexibility, and quality. When launched, it had already achieved its "steady state" cost target and profit margins were higher than traditional for that business. Senior managers at Company B credited their success to early collaboration among specialists in marketing, engineering, manufacturing, and testing. Manufacturability and testability was built in at the design phase and the manufacturing participants influenced the selection of product technology. Very few subsequent design changes were required. To accomplish all this, Company B had to achieve remarkable changes in the cultures of both its engineers and its manufacturing experts for this was a very different new product introduction process from that to which they were all accustomed. Participants gave credit to the strong new manager who was assigned to run this project for his initiative in establishing a positive attitude and momentum, thereby discouraging potential resistance to the sudden change. One observer in the company likened the product development team to a "daytime family." Company B is now attempting to sustain and incrementally improve the design-manufacturing interface in subsequent new product introductions.

Company C was well-established in its market for electro-optical equipment when it was suddenly leapfrogged by a competitor who introduced a better and less-expensive product. Though it was still fairly young and small, Company C had enjoyed rapid growth over the prior few years and was proud to be a leader in its industry. Their success had been based on allowing product engineers to retain a dominant role in defining the scope and timing of technological innovation in new products. In particular, the timing of development and manufacturing ramp-up were thought to be secondary to getting the best technology into the product. Now Company C had to try to respond in a more time critical manner. They could not afford the extensive process of engineering change orders (ECOs) that had characterized past product introductions, because responding to all of these ECOs delayed manufacturing's "ramp-up" to full production volumes. The company organized a special task force and made a conscious attempt to have product engineers and manufacturing experts work together from the start on a more equal footing. Product designers had to get the early involvement of manufacturing experts (and software developers) before designs were "frozen." A new emphasis was placed on designing parts to use existing castings from previous products, thereby saving tooling costs; prior to this design engineers thought that their job was always to design totally new parts. Following these procedures, the ECO syndrome was largely eliminated. The new product was designed, developed, and manufactured in a shorter period of time. Although quality and market acceptance was reasonable, it was not good enough to overcome the competitive threat that already existed. One key manager summed up the experience as follows: "We have lowered the wall between engineering and manufacturing, but we have yet to remove it." Company C attempted to learn from this experience by establishing a temporary task force to review this new product introduction. Shortcomings were noted and lessons for the future were formulated. The company is now trying to improve its new product introduction process even further, with full recognition that there are still cultural barriers to be overcome.

These three vignettes illustrate typical experiences of companies attempting to build or change the culture within which they design and manufacture products. Facing its own start-up, the founders of Company A built the desired culture from the outset. This was accomplished through the deliberate recruitment of competent technical specialists and managers all of whom wanted to work in a dynamic, integrated problem-solving climate devoid of traditional functional barriers. Company B, already established in its industry, launched a special project with a more interactive design-manufacturing interface. By so doing, they were able to turn a pending plant closing into a growth situation. Company C tried to do the same, although in a more informal and incremental way. Senior managers at Company C feel that their success in this instance was only partial and they are still working hard to avoid rigid functional approaches to new product introductions.

Each of these manufacturing companies grappled with a standard set of challenges in their continuing quest to build a more salable product.

THINKING ABOUT WORKPLACE CULTURE

Stanley Davis (1984) has written about corporate culture from the perspective of how to manage it. He makes a fundamental distinction between "guiding beliefs" and "daily beliefs." Guiding beliefs provide principles at a rather philosophical level; they deal with the best way to compete and how to manage a company. Such guiding beliefs, which rarely change, give direction to daily beliefs, which are more situational. Davis refers to daily beliefs as "the survival kit for the individual." While guiding beliefs support the formulation of corporate strategy, daily beliefs support the implementation of strategy. Davis uses the term *daily beliefs* to encompass how individuals are expected to behave with respect to innovation, decision making, communicating, organizing, monitoring, appraising, and rewarding.

Functional Silos

A central aspect of the workplace culture for those involved in product design and development is the amount of autonomy that individual, and groups of, engineers ought to have. Traditionally, large U.S. corporations have tended to develop what has become known as *functional silos* around their various specialized technical tasks. According to this vertical imagery, specialists work on small, contained projects, which are integrated through a hierarchy of managerial coordination and review. If technical solutions are acceptable to other specialists and managers, new assignments are made. If they are not, the problem is escalated to a higher level of command and control, where other specialists can eventually add their perspectives in a forum that tends to pit one organizational group against another. In many modern, complex electronic products, for example, hardware experts may propose that a particular mechanism be designed in such a way that the associated software demands, to be developed by another design group, are unacceptably burdensome. Or perhaps the proposed mechanical design contains so many parts that the manufacturing engineer is concerned about the cost to assemble the various components.

Professionalization and Specialization

Now consider what happens at a company that employs a variety of professionals, such as engineers. Here Joseph Raelin (1985) points to the dual problem of overspecialization and overprofessionalization and its associated challenges to corporate management. Overspecialization is a common characteristic within groups of design or manufacturing engineers, because each "field" is composed of a number of well-accepted subfields. For example, electrical engineers may be computer, power, or control engineers. Overprofessionalization, in turn, is characterized by an exclusive orientation to one's special skills or knowledge apart from the broader goals of the organization—that is, the proverbial "wall" between the worlds of product design and manufacturing. Despite any attempts to broaden the range of capabilities of its engineering work force, the managerial challenge inherent in dealing with long traditions of specialization and professionalization remains—that is, finding effective ways to pursue current business goals in part through the efforts of professionals with limited perspectives who conduct specialized tasks using technical skills.

Professionalization in design and manufacturing has traditionally meant that individual specialists desire and expect freedom to do the work that they know how to do best. Regardless of their own particular specialties, engineers as individual professionals are liable to share the common cultural norm of respecting others for their own unique contributions.

Although they are likely to have opinions as to the relative competence among members of another specialization, most individual engineers can be expected readily to acknowledge the importance of those who work in specialties other than their own. Of more personal concern than the number and variety of other specialties, then, is the work that any one specialist gets to do. As pointed out by Badawy (1971) engineers are normally highly motivated to work on important problems and probably will be frustrated to the extent that their company does not adequately use the professional skills and experience that they have acquired.

Such aspects of professionalism must be viewed in the context of workplace culture if we are to begin to understand the challenges for management in promoting effective product design and manufacturing. Rigid, functionally dominated organizational structures naturally lead one group of engineers to feel superior to another group based on the perceived importance of their collective contribution to the company's well-being. Everyone's organizational base becomes, in a sense, the center of the universe. Accordingly, it is common for product design engineers to feel that they are in a more elite part of the organization than are the manufacturing engineers. At the same time, manufacturing engineers often feel that it is their group that deserves the most credit because without them the product probably could not be made.

Conflicting Functional Goals

A related source of problems arises when professionals who need to interact to get their own work done have goals that conflict with each other. In his study of three business units of a major chemical company, Ginn (1983) demonstrated this relationship to be significant at the R&D-production interface. The production system is mainly oriented to achieving optimal rates of production output, while R&D seeks to introduce new products. Because the commercial introduction of new products interferes, especially in the short run, with achieving maximum production output, a classic case of conflicting goals often exists. Ginn and Rubenstein (1986) report that the interdepartmental conflict between these groups and the corresponding effort to place (and avoid) blame were relatively high. A classic struggle of this sort is to argue, most likely through memo warfare, whether a design engineering group's major proposed engineering change order will improve a product's performance enough to justify a lengthy delay and sizable expense for the required retooling.

The workplace culture of engineers is liable to be a blend of the

dominant values and norms of their profession coupled with cultural elements created through time by the management of the particular organizations in which these people work. In particular, formal reporting and
review relationships, combined with criteria for the successful completion of a job, will strongly influence the type and intensity of "working
together." Managers of engineers may attempt to encourage professional
staff to work together, yet the scope and span of such collaboration will
be limited by those individuals' background, education, and training, as
well as by any natural boundaries created by the organization of work in
the company. Performance measures that focus on the optimization of the
components of a system (i.e., a product or a manufacturing process),
rather than on the system as a whole, is another common catalyst for
specialization.

Challenges in Developing a Conducive Workplace Culture

A synthesis of the organizational and management literature suggests that
the following three challenges need to be met in establishing a workplace
culture conducive to the effective introduction of new product.

1. **Achieving Strong Collaborative Behavior among Specialists
   with Different Backgrounds.**   The goal is to create a daily belief that internal collaboration is better than either competition or
   independent, partial problem solving. The challenges include: (a)
   achieving an effective blend between science and experience; (b)
   making technological complexity a catalyst to achieving integration, rather than being a formidable barrier; and (c) establishing,
   for a new product, a common vision that promotes constructive
   modes of analysis and conflict resolution.

2. **Balancing Firmness and Flexibility in Bureaucratic Procedures.**   The goal is to provide a culture which supports simultaneous attention to innovation and efficiency in the design and
   development of new products. The challenges include: (a) instituting design deadlines without overly constraining engineers
   from doing their best work; (b) encouraging the introduction of
   new technology without missing strategic market windows; and
   (c) developing project managers who can combine the skills of
   leaders and cheerleaders.

3. **Providing a Positive Internal Market Mechanism.**   The goal

is to establish team objectives to be shared by the various participants in new product development. The challenges include: (a)
establishing performance measures that create team-based incentives for professionals used to independent assessments of their
work; and (b) encouraging participants to help create new work
settings that differ radically from that to which most have already
become comfortable and proficient.

Patience and perseverance in dealing with these tensions is required,
for they are complex and there is no easy "cookbook solution" to fit all
situations.

ROOTS AND TRADITIONS OF SEVERAL KEY PLAYERS

With these perspectives on workplace culture and the nature of professional work, we turn to several types of specialists whose work is central
to successful product design and development, such as design and manufacturing engineers, industrial designers, and human factors specialists.
This section provides a sense of who these specialists are and what their
traditional orientations and workplace cultures tend to be. With this rather
concrete orientation to the variety of players, we are better able to appreciate the challenges in developing a conducive culture for joint problem
solving on NPI teams.

By choosing only these particular players for coverage here, we do
not mean to minimize the important roles played by other fields represented on NPI teams—notably, marketing, purchasing, and field service.
Other chapters of this book indicate the nature of their roles in the process
of product design and development.

Design Engineers

Effective product design, which essentially consists of problem-defining
and problem-solving activities, needs a managerial environment that
guides and facilitates such work. Two key aspects of this culture are the
handling of design deadlines, and dealing with technological complexity.

Despite its creative mood, product design needs to be managed. Mature companies often employ top-down decision making to create stability
for those who do product design. Such an approach requires that dead-

lines for producing designs be set in advance and that those deadlines be taken very seriously. This emphasis on time-based deadlines creates a natural pressure for design engineers to attempt to modify existing designs without the kind of thorough analysis and testing that would normally be desirable. In their study of a large design engineering unit in a major U.S. automobile company, Liker and Hancock (1986) observed that one result of this approach was that extensive and unanticipated resources had to be allocated to the subsequent correction of problems. They also noted the "vicious cycle" that arose when future new products were neglected while scarce time and resources were applied to fighting the fires from former faulty designs.

The important point here, from the perspective of managing the design-manufacturing interface, is that in some working environments, design engineers have become accustomed to meeting what they perceive as being serious though not critical project deadlines. The deadlines are serious enough that one does not wish to fail to have a product design ready at the preestablished time. Apparently, organizationally imposed criteria, the "timely handover" of a design, can dominate the professional norm of producing a high-quality design. What makes this compromise possible for the design engineer are two other elements of the work culture: that it is acceptable, and even expected, that change orders be issued to "fine tune" the design after its release; and that it is usually someone else's responsibility at that point.

This approach to product development can be functional for the design engineer in that there is a built-in opportunity to "look good under pressure" on two occasions: first, when the design is released; and then again when the fine tuning is accomplished. To the extent that design engineers are rewarded for such behavior, and the rest of the organization finds it to be acceptable practice, these engineers are not likely to instigate a radical change in the process of new product introduction. Nor is a top-management statement about the importance of "doing product innovation right" likely to alter the traditional patterns of behavior. The design of shortcuts to meet schedule deadlines is usually needed. In addition, performance measures must be instituted to stress early, effective use of cross-functional teams, rather than relying on subsequent engineering change orders and cost reductions.

Companies that would prefer to achieve "mature first costs"—where a complex product is designed so that it can be manufactured as intended from the beginning and at the target cost—must take appropriate steps to overcome the culture that naturally resists this objective. This kind of planning and scheduling can be particularly difficult to accomplish where there is extensive specialization in the design of a complex product. Here, the aggregation of component designs into system designs is an independent activity and the release of the full product design must be carefully monitored and controlled.

The technological challenges facing design engineers, given their base of knowledge, vary greatly from industry to industry and from company to company. Each design engineer has a personal repertoire based on prior experience in defining and solving particular technical problems. In each company engineering solutions are in part constrained by the scope and content of existing data bases. One should not be surprised to find that the extent of technological complexity inherent in traditionally designing a certain class of product in part determines the culture of that organization's design function.

Consider the following illustration. In large complex organizations with a relatively stable and well-understood product line, engineers often work in a hierarchical structure. Supported by experienced draftsmen, and engineers in supplier firms (and consultants when needed), those with the title *design engineer* may not require extensive technical knowledge. Liker and Hancock (1986) found that such engineers built experience mostly in "coordination of information flows and policing engineering changes," rather than in developing new, effective design solutions. These researchers observed that the culture of such an organization is likely to resist the development of specialized technical competence. Instead, in these types of companies, ambitious young engineers are encouraged to concentrate instead on behavior that will lead to a rapid promotion out of product engineering and into management.

The same study also found that smaller companies tended to produce state-of-the-art products and built a strong record of success in satisfying the requirements of their market niche. The companies traditionally have other distinguishing features in their product design culture. They are apt to be engineering-driven, as in the example of Company C (discussed above). Here, design engineers often have the power to imagine the next technological breakthrough and tell marketing managers what they will provide and approximately when this new product will be available. It is not unusual, in such settings, for design engineers to incur little or no penalty for missing their deadlines. Designers and their managers in such settings are likely to be sympathetic to the notion that product design is a creative process which is inherently hard to forecast.

Manufacturing Engineers

Traditionally, manufacturing engineers have been comfortable with the following prescription for their work: take instructions, work from a complete design, and then get things done! The traditional "handoff" from design to manufacturing is like the beginning of the last leg of a relay race. The manufacturing team is the last runner and will try to make up ground in the sense of figuring out how to produce the product as currently designed, how to adjust manufacturing processes in response to product design changes, and how to "ramp-up," as rapidly as possible, to commercial rates of production. Establishing a workplace culture that supports such efforts is a challenge unto itself.

To some extent the size of the company, as reflected in the scale of operations, is inclined to affect its manufacturing culture. Company size brings other important variations, including the extent of resources and support for ongoing professional development through company-sponsored training programs or tuition remission plans for employees pursuing advanced degrees. Additional factors that tend to vary with size are the structure of the manufacturing organization, the number and type of manufacturing engineers, and the balance of power between corporate staff and plant level personnel.

In large companies, manufacturing engineers may work in groups of similar specialists where narrow technical competence is valued and rewarded. Such specialization brings the possibility of more sophisticated solutions where manufacturing engineering problems are naturally decomposable. However, it also brings the risk of unproductive lapses in communication and coordination when technical problems are more interrelated. The existence of independent groups of specialists also tends to generate "turf battles," particularly over issues of resource allocation.

Advanced manufacturing groups are also common at the corporate level in larger companies. Members of such groups often have advanced degrees and relatively high salaries, thereby tending to set them apart from manufacturing engineers in individual plants. The corporate groups tend to be especially familiar with the latest computer-based automation technologies, although they may lack current, in-depth familiarity with the manufacturing capabilities of particular plants. Typically, they function as internal technical consultants to the plants in times of modernization, capacity expansion, or major new product introduction.

In such instances, bridging the cultures of the corporate and local manufacturing engineering groups can be a challenge unto itself. Larger companies with this structure need to leverage the investment they have made in advanced skills at their corporate groups. Plant-level groups may resent that this investment was made in the first place. Stakes can be high for the company, in terms of both return on investment and lost opportunity cost, if such efforts are not fruitful, because new products can come from the exploitation of existing manufacturing processes.

Smaller companies, in contrast, tend to have very few highly trained and experienced manufacturing engineers. They are more inclined to employ generalists at the plant level and to operate without a corporate advanced manufacturing group. Manufacturing engineers (whatever title they work by) tend to develop experience with particular sets of machinery, fixtures, tooling, or software. Frequently, personnel turnover in these positions is not high and much of the institutional memory about the manufacturing capability is in the minds of individual engineers. Documentation of prior projects may be spotty. Those who worked on some component of the plant may informally inherit the responsibility for all subsequent problems or modifications. In such a setting the mood is liable to be more pragmatic and less bureaucratic.

The above description relates to conditions common in the United States. A different scenario for manufacturing engineering is, however, possible. When one looks to the leading manufacturing companies in other countries, notably in Japan and Germany, the demographics and workplace culture of manufacturing engineers stand in stark contrast to the typical U.S. situation as presented. It is not unusual in such companies to find Ph.D. engineers on the plant floor. Moreover, in the Japanese computer industry it is common for top-engineering graduates to be reassigned between product design and manufacturing (Westney and Sakakibara, 1986). While there are no statistics comparable to those from the United States cited above, the work of R. Jaikumar (1986a) suggests that such radical changes in the role of engineers in manufacturing are both possible and necessary.

Consider, for example, the case of Hitachi Seiki, a large Japanese machine tool manufacturer (see Jaikumar, 1986b). In 1980, this company established an engineering administration department, consolidating the functions of machine design, software engineering, and tool design and stressing the more generic role of systems engineer. The head of engineering performed this reorganization to enhance the coordination among these functions.

The culture and skill base at Hitachi Seiki evidently supported this plan. A team of just 16 engineers simultaneously designed, built, and installed with successful results three different flexible manufacturing systems within an 18-month period. This tight-knit group of systems engineers completed these new product introductions (for their own internal use) in record time and consistent with the company's preestablished return on investment criteria. In summary, manufacturing engineering cultures in technology-oriented U.S. companies need to shift over time to be more like those at Hitachi Seiki.

Industrial Designers

The field of industrial design has traditionally been central to the introduction of new products in style-conscious fields such as home and office furnishings. In the 1990s, the importance of industrial design has been recognized in fields as diverse as computers, medical instruments, and industrial equipment. The background, skills, and styles of industrial designers are much different from design and manufacturing engineers, though they often work together on NPI projects.

Industrial designers attend educational programs that are traditionally quite different from the engineers with whom they must work. Industrial design programs are usually located in schools of art rather than sciences. The programs contain studio courses involving the construction of models, in the tradition of architecture, and considerable project work. The more rigorous industrial design programs are moving closer to fields of engineering and include courses in the theory of structures, human factors, and materials, such as wood, metal, or plastics.

The traditional contribution of the industrial designer has been to the overall shape, style, and appearance of a product. In the 1990s, the industrial designer has begun to play the role of the synthesizer in the early concept development phase of a new product introduction. While the many technical specialists involved in product design and development consider one or more particular aspects of a product, the industrial designer is trained to maintain a focus on the product as a whole. Industrial designers add value to the design process by making physical models of different versions of a product for consideration by others and by raising ease-of-use concerns throughout the design sequence. Industrial designers are particularly concerned about the perceived quality of a product in the minds of users. This dimension of quality incorporates both the utilitarian and the symbolic functioning of products in their context of use. One example of this perspective is that the design of a chair is finished only when someone sits in it.

The industrial designer is trained to ask questions about products, not as ends in themselves, but rather in terms of their use and understanding by people in everyday life. Industrial designers, therefore, try to balance issues of product performance with cultural questions about the context of use. The look and feel of a product are the primary domains of the industrial designer. As products increasingly contain aspects of software (information systems and programs) combined with hardware, the interplay between hardware and software becomes more complex (Heskett, 1989) and the challenges for industrial design are accordingly enhanced. The classic Bauhaus school of industrial design insists that a product's form follow its function (simplicity), while others strive for more distinctiveness (colorful, whimsical, fun). Regardless of their orientation to matters of style, industrial designers typically work to develop products with "design integrity," an overall image that is compatible with a company's culture and its customers' expectations.

Industrial designers may work on an NPI project either as an outside consultant employed by a design firm or as a member of an in-house design group. Companies such as Sony, IBM, and Motorola, while occasionally using design consultants, have developed their own staff of industrial designers who are familiar with the company's engineering, manufacturing, and marketing capabilities. In 1990, the American automobile manufacturers had the largest staffs (as many as 600 designers). While a large global company such as IBM may have about 40 industrial designers working on new products, an in-house team of several industrial designers typically serves a sizable business unit. Most manufacturers do not have an industrial designer on staff and, if they want to use this resource, they hire an independent design consulting firm.

Integrating industrial designers within an NPI team has its own set of challenges. Industrial designers, as stated previously, are generalists and do not overlap in their background or skills with engineers who are called *designers*. In fact, their orientations can be so different that the one word design can itself hinder communication, since it means different things to different people. In addition, many industrial designers pride themselves on their skill in getting to know what the customer needs. Here, there is a natural potential for confusion between what a marketing specialist identifies as "customer requirements" for a new product and the inputs provided by an industrial designer. In both of these cases, the unique contribution of the industrial designer is the ability to visualize what a product might look like to the customer.

Because visualization is so important in the early phases of an NPI project (Phases 0 and 1), the industrial designer is particularly valuable at this time. In fact, some companies, such as CibaCorning Diagnostics, use industrial designers as project leaders during this stage to allow them to synthesize the contributions of the various functional specialists into sketches and models of different product design possibilities. These models are increasingly being produced on advanced CAD (computer-aided design) systems, but the traditional approach is to create physical models from wood, clay, or plastic.

Visual, as distinguished from abstract, models can contribute greatly to the cross-functional communication so essential in effective product design and development. As industrial design models progress from basic concepts to product specification, all other functional specialists need to get involved. An appealing external shape needs to be assessed in terms of whether it can be machined or molded at the necessary levels of quality and cost, and also whether the product's "insides" (i.e., circuitry, tubing, or other components) will fit. A proposed external material or color (or shape), may also be assessed in terms of such considerations as wear, safety, and market appeal. Good industrial designers will try to consider all of these factors in making their models, but others in the company will probably have their own opinions.

Much is at stake in the effective use of industrial designers because they are inclined to be more oriented to the notion of "product integrity" than most other contributors to product design and development. Product integrity, although difficult to define in advance, becomes a critical ingredient of customer satisfaction. Some successful industrial designers see themselves as promoting product integrity by facilitating an integrated design process. Others stress the value of their creative inspiration in ensuring product integrity. Either way, the NPI project manager must strive to work the industrial design perspective into the team's problem-solving activity so as to build on its strengths, while recognizing that others also need to make essential contributions early on. Managing this process can be especially challenging when an outside industrial design firm is being used for the first time.

Human Factors Specialists

The study of *human factors*, often called *ergonomics,* is also of growing importance in the design of a wide variety of products. Ergonomics is a combination of psychology, physiology, biomechanics, anthropometry, and experimental design. The human factors specialist usually has a graduate degree and is essentially responsible for representing the user population in the product development process. They often work in industrial design groups and contribute their specialized skills to the overall design of a new product.

By studying how products are actually used, the human factors specialist can assess different design features and functions in terms of the product's ease-of-use, safety, and reliability. The work of the human factors specialist is highly structured. It is based on careful observation and the use of logic, often rooted in prior studies and data. This approach can be valuable at various phases of the NPD process in the refinement of product concepts, the development of design criteria, placing constraints on design directions, and testing design solutions.

Human factors specialists often work closely with industrial designers, since the shape and feel of a product and the location of its external controls may have direct physical or psychological impacts on the user. For products such as chairs, desks, personal computers, industrial, or medical equipment, the ergonomics aspect of design can spell the difference between a good or bad product. There may be conflicts between the ergonomics point of view and that of a creative stylist. A successful industrial designer will easily recognize the importance of questions of ease-of-use and many industrial designers have some training in methods of ergonomic analysis.

With the advent of the personal computer and engineering workstation, many human factor specialists now work closely with software designers. Their role is to review screen designs and user options. Psychological and physiological impacts of using software interactively has become an important design criterion and it is emerging as a specialty in the field of ergonomics. Software applications are often developed with early prototypes that can be tested from a human factors point of view, even before the full set of software capabilities is designed. A related contribution of ergonomics is the development of self-taught interactive electronic "instruction manuals" to replace frustrating reams of printed manuals that many factory workers had to rely upon for guidance in their everyday work.

Ergonomics specialists often struggle to represent the user in the face of increasingly complex consumer products. One need only think of the trends in wristwatch and VCR design to understand the difference these specialists can make. As marketing pushes for more features in such products and engineers figure out how to increase product complexity at low-incremental cost, the result may be an electronic watch that takes dozens of steps to set or a VCR that requires a training course to operate. The ergonomics specialist (along with the industrial designer) will resist this trend by working to simplify the user interface without necessarily reducing the product's functionality. Human factors specialists even employ methods for testing product prototypes directly with potential users with less-than-average cognitive or dexterity skills. In such instances, they view their contribution as the early detection of a product that is designed to inhibit successful hands-on use.

One of the complicating factors in using human factors specialists effectively is that they lack the traditional functional power base of other participants in product design and development. Project managers must therefore make a special effort to bring such resources into the team in ways that promote their effectiveness. The best way to do this is to emphasize from the start the importance of customer satisfaction in all aspects of product design and development. When the quality targets for a new product explicitly reflect ease-of-use, the human factors specialist will be better positioned to gain an equal voice in design meetings. Conflict cannot be expected to disappear, however, because the advice of the human factors specialist may still conflict with that of marketing, design engineering, manufacturing engineering, purchasing, or field service. One cannot fully escape the inherent trade-offs among the various quality and cost targets for a new product.

ACHIEVING COLLABORATIVE BEHAVIOR

The general subject of collaboration within working groups is central to the field of organizational behavior and is well-covered elsewhere (e.g., in the work of William Kraus, 1980). The widely accepted "solution" of building more effective teams has been developed to the point that general-purpose diagnostic instruments have been available for some time (Francis and Young, 1979) and generic prescriptions are in the management literature (Wolff, 1989). However, with rare exceptions (e.g., a recent study by Hans Thamhein, 1990) this literature does not focus on new product introduction.

To begin to address the challenges in developing team solutions for enhanced product development, two types of barriers need to be examined: (1) how people think (cognitive limitations), and (2) what they do (behavioral patterns) as they participate in their company's efforts to introduce new products. Effective cross-functional collaboration requires that the various NPI team members communicate well with each other and can develop shared views of problems that need to be resolved. Such cognitive compatibility is liable to be lacking in most U.S. companies, as has been discovered in our own research and elsewhere. Kathleen Gregory (1983) has argued on a conceptual anthropological basis, that forming a "culture" that is truly conducive to continuously improving NPI is severely limited by "native-view" paradigms that create culture conflicts in organizations. In a similar vein, Deborah Dougherty (1989) has documented the existence of interpretive barriers to successful product innovation. This means that subgroups with different occupational or divisional cultures inevitably bring to organizational interactions their own meanings and senses of priorities (i.e., "thought worlds").

To the extent that this situation exists in any particular company, knowledge creation within the NPI process will be constrained. Martin Ginn (1983) showed how conflicting goals between R&D and production specialists in a major chemical company led to conflict and counterproductive behavior in the process of product development. Gupta, Raj, and Wilemon (1985) surveyed the R&D-Marketing interface in 167 small- and medium-sized research-intensive firms and found that inherent communication barriers and insensitivity to the others' capabilities and perspectives were significant barriers.

In summary, there is ample evidence that companies with a tradition of sequential, functional-oriented product development have deeply ingrained routines in which the work of any single department has a certain internal "wholeness" (using Dougherty's term). Organizations lacking integration may be comprised of members acting from numerous internally consistent (within marketing, R&D, manufacturing) but externally conflicting cultures. A successful NPI process requires individuals to break

FIGURE 4–1
Factors Affecting Successful Collaboration

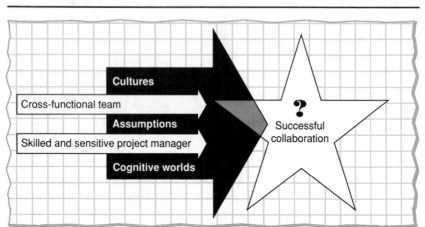

out of such routines and collaborate across functions in the pursuit of new knowledge, coherent decisions, and enhanced product outcomes.

Behavioral patterns, as well as cognitive frameworks, thus affect the nature of problem solving in the course of new product introduction. The new type of communication and interaction will seem unnatural to many NPI participants accustomed to a more circumscribed and insular role.

Managers cannot expect simply to eliminate these fundamental problems in communication through rapid restructuring of the organization. Even when a skilled and sensitive project manager is in charge of a suitably cross-functional team, the probability of successful collaboration will also depend on the mix of cultures, assumptions, and cognitive worlds (Figure 4–1). More deliberate efforts at organizational design, following the sociotechnical systems approach, has led to identifying those structural changes that will also improve the workplace culture. In the case of Zilog, Inc., as described by Taylor, Gustavson, and Carter (1986), a more conducive workplace culture for engineers and designers was created. While certain structural changes can promote collaboration, implementing them within most companies is far from an easy task.

Although the formation of cross-functional teams is now a generally accepted strategy for pursuing collaboration from the start of an NPI project, such action may not be sufficient. Even when organizational integration is pursued, cultural isolation, traceable to the cognitive dilemma of independent "thought worlds," may persist. Common traps in group problem solving can easily arise. First, each coalition may take its own position for granted and not offer it as a starting point for further discussion. Even more subtle, NPI participants may assume that their own natural meanings and priorities are shared. Conflicts based on confusion over language, often remain tacit and lead to misunderstanding and wasted time and effort. Conflicts based on limited views of problems and outcomes often end in irreconciled debates. The occurrence of such cultural conflicts is somewhat unpredictable. They may arise only in a few situations or in many and may have different outcomes.

When one begins to acknowledge the different roots, traditions, and styles of the various participants in an NPI project, it is not surprising that true collaboration may appear to be elusive. Progress along these lines is certainly possible but elimination of the associated difficulties is probably unrealistic. Senior management should encourage, through the formulation of team objectives and assignment of personnel, the daily belief that internal collaboration is better than either competitive interpersonal behavior or independent and partial problem solving. To do this effectively, management must be aware of the forces that favor such action, the natural impediments likely to arise, and the role that the project manager can play in supporting such developments. These topics are discussed in the sections that follow.

Favorable Forces

Despite the challenges identified above, managers can take advantage of certain forces that naturally favor the creation of the kind of workplace culture so important to effective NPI.

Start-ups have a special opportunity to establish adequate collaboration at the outset.

Three favorable factors can combine when a new company is engaged in its first new product design and development: (1) a cohesive vision; (2) a compatible group of experts; and (3) a highly interactive problem-solving environment. These factors may exist for the following reasons. First, the founder of a start-up company normally has a coherent product vision that provides a focus for all key participants. Second, start-ups have no counterproductive cultural traditions. The key company participants, if not cofounders of the company, will be hired directly by the founders. These people will be the core of the new product development team. They can be screened and hired for their interest and ability to work in a cross-functional environment. Third, start-ups are, relatively speaking, small organizations. The natural smallness of their initial NPI team facilitates informal "group" decisions based on a marketplace of ideas. The challenge for start-up companies is to maintain these natural advantages even as the company moves to introduce its follow-on products.

For companies already in business, the existence of a crisis atmosphere, or at least an awareness of prior avoidable NPI failure, helps to a degree.

The power of a crisis in stimulating organizational change has been documented for a long time and across many industries. In fact the popularity of the word *proactive* is no doubt a response to the common but unfortunate organizational condition known as being reactive. In the field of new product design, as in so many others, the reactive mode has been very common.

Consider the situation when a technology-based product fails in the marketplace and the business unit that introduced it quickly falls upon difficult times. We often find, under these conditions, that the various functional experts working on the next new product design share a common interest in collaborating more closely with each other. Under such a condition, it is not surprising, for example, for manufacturing people to work more aggressively than usual to ensure that their perspectives are acknowledged early in the NPI cycle. This type of collaborative behavior is most likely to occur, given this gloomy scenario, when a strong product champion changes the workplace mood to a more positive direction. Low morale, even under crisis conditions, is not a strong motivator.

Prior failures in the process of product design will lead many companies to try to understand their competitors' better abilities along these lines. Benchmarking exercises of this type seem to be more common after a company has lived through an especially obvious failure on their own part. Another increasingly common catalyst for change is when a parent company demands shorter product development times from one of its shaky business units.

A preoccupation with existing products can hamper NPI collaboration, while lack of such distractions can help.

One of the most basic sources of distraction, in NPI and all other business activities, is derived from the existing performance measurements in effect in the organization. Manufacturing performance is traditionally measured in terms of meeting current production schedules, and the quality and cost targets for existing products. All of these measures may tend to be adversely affected when key manufacturing personnel and facilities are allocated toward the introduction of a new product. In this scenario, it should not be surprising to find that new products are considered to be distractions from the higher priority existing products. This situation is exacerbated even further when production sources for a new product already have fully utilized capacity. Here, under the same traditional (and short-term) measures of performance, manufacturing management may have little incentive to collaborate with an NPI effort by assigning scarce process engineering talent to work on it.

Conversely, some of the positive examples of early collaboration on an NPI team may be traced to the lack of distractions from existing product delivery. Most manufacturing engineers have ongoing assignments at either an advanced manufacturing facility or a high-volume plant. When demands on their time from these traditional sources are lighter than normal, these engineers are more available to get involved early in the design stage of a new project.

Other organizations naturally avoid this potential conflict of interest simply because new products always represent something close to a "bet-the-company" situation. At Amdahl Corp., for example, there may be genuine disagreement about whether any particular idea for a new mainframe computer product is worthy of pursuit. Nevertheless, once an NPI is under way the culture of the company is such that it receives top priority from all who are involved.

Finally, many larger companies have adopted the approach of a temporary, full-time co-located NPI team isolated from ongoing operations. Called *skunkworks* or *tiger teams*, these small teams of handpicked experts are usually given special up-front authority to take a new product idea through the conceptual design phase. As the project shifts to more detailed design and development, the original team is enhanced as needed and the independent nature of the product will be retained until the hand-off for volume manufacturing.

Management must set the tone for collaboration.

Perhaps the best single tone-setting action by senior management is the establishment of a broad and continuing corporate goal of quality improvement. Such a shared goal promotes the spirit of collaboration within NPI teams and between the team and other related groups in the company. A shared value to pay careful attention to customer needs can create a common vision for the team. A specific and coherent challenge issued by a senior executive to the newly formed team will also set the appropriate tone for enhanced teamwork. Connections between effective product design and development and ongoing programs of total quality management (TQM) are discussed in Chapter 11.

In addition, trust between manufacturing and engineering improves when a project manager, who believes strongly in cross-functional teamwork and has the interpersonal skills to match this belief, is appointed. This feeling is further enhanced when there is a product champion, with a track record for being innovative, who clearly has the trust of top management. Promoting the use of design-for-manufacture techniques, helping participants understand the entire project, and a corresponding company commitment to training all contribute toward meaningful collaboration across NPI functions.

Management sets the stage for collaboration by setting clear goals and then empowering the team to identify and solve the problems of product design and development. Under the phase-gate system of management control (Chapter 2), all of the formative and detailed decisions are made by the team (and are subsequently reviewed). If those on the executive review committee can refrain from introducing new design options or imposing their personal preferences during the later phases of the project, they will have contributed greatly to the spirit of collaborative teams.

Finally, when faced with a "bet-the-company situation," top-management support at the beginning of the NPI project, along with the stated intention to provide all required resources, will be a critical catalyst to collaboration. It is easier for collaboration to naturally occur in companies that "live or die" with their next product.

Co-location helps ensure early bonding, greater productivity (improved coordination and streamlined decision making), and more informal contacts that promote creative thinking. All this ads up to more effective product designs and the type of learning that supports future NPIs. Unfortunately, however, full co-location often is infeasible in large global organizations, and temporary measures must be taken.

The obvious theory behind co-location is that people who work in close proximity with each other are much more likely to communicate with each other in an ongoing, informal as well as formal, basis. Collaboration is much more natural under such conditions. Some companies speak of their NPI teams as "daytime families" under such conditions.

Co-location, then, is a good idea if it is indeed a feasible alternative. A start-up company often has the luxury of being able to co-locate all NPI participants, at least until the first plant is built. In large organizations with widespread facilities total co-location of an NPI team may be impractical. Here, management should consider the value of taking steps toward some partial co-location. For example, it may be possible to co-locate the engineering participants of an NPI, even if they cannot be brought to a common location with marketing or manufacturing. At least then one can expect better collaboration among the various product design specialists. Co-location benefits require that participants work in very close proximity to each other. Being in separate buildings, even in the same community, is not enough. Benefits in collaboration have even been reported from companies that have brought together, into a single workspace, NPI participants who were formerly assigned to different wings of the same building.

Sometimes co-location will be deemed infeasible due to the prevailing organizational structure. Functionally oriented, technology-based companies sometimes turn to a matrix structure as the basis for their NPI efforts. Such companies group their technology specialists together, and if the company is large, there may be considerable geographic dispersion among the groups participating in the same NPI. This is even more of a problem for a global company with engineering and manufacturing facilities in different countries. In this situation, some sharing or splitting of design responsibilities across thousands of miles will frequently be necessary. Substitutes for co-location often need to be pursued aggressively if true collaboration is desired.

Collaboration will be enhanced if the project team spends time at the beginning to learn from prior NPI mistakes and successes. So doing will be facilitated if a thorough assessment of prior NPIs has been conducted and widely distributed.

A formal postmortem is not always necessary. For example, sometimes the problems with the prior product introduction were so serious and visible that they are already apparent to all those participating in an upcoming NPI. Usually, however, a formal NPI project assessment (often called a postmortem) can provide the type of participation, investigation, and documentation that promotes considerable organizational learning for future NPI projects. How to do this effectively is described in Chapter 13.

A special situation arises when the product life cycle is such that there are many years between successive major NPIs. Then, it can be harder to sustain an institutional memory and the avoidance of key staff turnover becomes especially important. The shortcoming of NPI project assessments in this setting is that the value can be diminished when economic, competitive, and technological conditions differ widely from one major NPI to the next. The value of a project assessment in enhancing NPI collaboration is also reduced when attempting to transfer lessons from one product line or market segment to another.

A general spirit of collaboration will be facilitated by an early, customer-oriented, shared vision of the objective of the current NPI. There are many different ways through which a common vision may be achieved.

At Amdahl Corporation, for example, a strategy of being second-to-market with price/performance leadership, requires a rapid response to the market leader's product announcement. Everyone in such a company easily understands this overriding requirement, and this helps promote rapid consensus on the new product vision.

For some companies, basic performance and availability requirements for some products, such as a new jet engine, are tied to the requirements of other related products (e.g., airframes) and therefore provide a shared NPI vision. In companies that are highly regulated, the government agency becomes, in effect, an intermediate form of customer. The importance of receiving required approvals (e.g., FAA certification) is understood by all, thus promoting a common view of the proper NPI process and the desired interim outcome. In a more abstract manner, when an industry has developed a traditional set of milestones in introducing a new product, all participants in the NPI can quickly see how their efforts fit into the larger effort.

Sometimes a common vision for a new product comes from market feedback, either direct or indirect. One manufacturer of industrial shelving experienced this phenomenon when a key distributor provided a crucial function by collecting customer complaints and helping the manufacturer to appreciate the full range of product improvements that would be desirable.

Tools such as electronic data bases and integrated CAD/CAM and other special devices can facilitate collaboration if the participants are so motivated.

This use of design-based technology can help suppliers as well as internal company participants to communicate with each other. This subject is discussed in more detail in Chapter 6. One low-tech device that has long been used to facilitate joint problem solving has also proven to be valuable—a blackboard for recording ideas. Occasionally, the timely use of a neutral technology expert can resolve a significant design dispute.

Natural Impediments

Sharply differing philosophies among functional groups, and perceptions that these groups work differently (e.g., scientific versus experiential orientations), will hinder NPI collaboration.

For example, when marketing and manufacturing are trying to solve problems with an existing product, they are more apt to see NPI as being an enhancement of that product rather than a new and different enterprise. Many product engineers, meanwhile, naturally prefer working on entirely new designs and technologies. In most companies, design engineers and advanced manufacturing engineers will naturally relate more to each other's perspectives than they do to those of a project manager with a marketing background. However, when responsibility for new product development is in the hands of a corporate design engineer, there can be some difficulty in getting adequate collaboration with manufacturing and marketing specialists at the plant level.

Other natural barriers to collaboration also can be expected to surface. Some examples:

When the NPI is not a top priority, continuity in the cross-functional team approach can suffer.

Natural suspicions can get in the way: for example, reluctance by commodity managers to accept parts that are qualified abroad, or a fear that one's own job is being threatened when one's traditional design/development tasks are given to other groups.

Manufacturing pays most attention to current high-volume products.

Company restructuring becomes a distraction.

Collaboration is impeded when global NPI involves different languages and nationalities.

When organizational objectives and performance measures for the various functional groups are not cohesive, there is a tendency for parochial behavior.

The project manager and team must deal with problems of underestimating the challenges faced by participants with different functional responsibilities and experience levels.

The engineering culture is often characterized by perfectionism, while manufacturing is pragmatic (expedite to meet the schedule).

Borrowing a product engineer who is oriented to existing products, as a key part-time participant in a new product introduction can be less satisfactory than the use of a dedicated full-time new product development engineer.

The Project Manager's Role

As discussed in Chapter 2, effective product design and development requires a process that is disciplined but not overly regulated. A workplace culture in which firmness is offset with a constructive amount of flexibility is needed for the NPI team to develop a collaborative culture. While much of this balance needs to be reflected in the executive review process, on a daily basis it is the project manager who plays the most pivotal role in this respect. The project manager must act in ways that support team members' collaboration with one another.

Problem solving by the team can be encouraged, for example, through the style of the project manager; and the project manager needs to establish an expectation that the team is supposed to provide solutions rather than roadblocks. This will be easier to accomplish if the project manager has skills in planning and facilitating consensus among the members of the team. The NPI project manager should try to ensure that cross-functional reviews are frequently conducted within the NPI team.

Sometimes NPI team meetings become so large that they are unwieldy. It is then difficult to establish an effective balance between firmness and flexibility: excessive firmness discourages participation, and too much flexibility yields nonproductive discussion. A potentially constructive procedure in such instances is to look to technical supervisors in different disciplines, rather than a single project manager, to decide who ought to attend important NPI meetings. Whether such a gatekeeping function works to the benefit or the detriment of the NPI depends upon how such screening is executed.

The project manager can also be influential in obtaining cooperation from individuals and groups who need to be involved but who are not part of the ongoing core team. Negotiating skills can be important here but the perceived priority of the project will naturally affect the attention level of these extended participants.

A project manager who is granted significant authority for the NPI effort may become involved with the selection of team members. Obvious considerations in such decisions are the skills, availability, interests, and track record of potential team members. Less obvious, but important, are the problem-solving styles and personalities of the various individuals who are assembled on the team. Teams need a balance of styles that support the full range of activities, such as planning, data collection and analysis, and decision making, associated with product design and development. Effective teams will avoid dysfunctional behavior such as paralysis through analysis, unjustified persistence, premature decisions, endless exploration, or destructive bickering. Other personality factors of its members, such as enthusiasm, openness, tolerance for ambiguity, and thoroughness will affect the effectiveness of the team. Some companies are seeking to establish more team synergy by being more systematic in considering these kinds of factors when they select core team members. The others, who essentially ignore such matters, operate at a severe disadvantage.

Project effectiveness will also be affected by the authority granted to the project manager which, as mentioned in Chapter 3, will vary from company to company. Formal authority is just the beginning, however, and project managers must be skilled in using it. When issues of priorities or resource availability arise, all project managers must be ready and willing to seek backup support from senior management (i.e., authority from above). Negotiating skills are usually essential, especially when attempting to obtain cooperation from individuals and groups whose contributions to the NPI effort are required, even though they are not part of the ongoing core team.

Finally, the project (or program) manager can promote collaboration simply by ensuring that the product concept is carefully articulated and fully understood by all team members. This will help everyone to share a common overall vision for the project. This can be helpful even though the team members may have differences of opinion on how to translate the concept to particular elements of the product or process design.

The Use of Incentives

In most large companies incentives are needed to encourage individual NPI participants to shift from traditional functional orientations toward true cross-functional collaboration. With such incentives, key team members will work hard to become more adept at cross-functional problem solving, using their experience and training. Sufficient value must be placed on innovative new product development to guarantee shared priorities and common goals. The challenge is to create a set of incentives that will work, given the realities of the company's current situation and traditions. A special transition challenge is to encourage participants to help create work-settings which are radically different from those that, while less effective, are a more comfortable fit with their own prior education and experience. It should be noted that start-up companies have built-in incentives for a while because the equity interest typically held by all major participants provides adequate motivation to work toward a common end.

We have already suggested how symbolic incentives can work, even under adverse conditions. The creation of a "daytime family" spirit in an NPI team can lead participants to work toward a common end. A team faced with a challenging NPI assignment can develop a healthy "we'll show them" attitude that favors collaboration. Following an initial NPI success, highly visible praise by management can lead others to want to join that project team. Some companies have found that coining team slogans and publicly communicating the achievement of major milestones can motivate NPI participants to work together.

Personnel evaluations and other human resource development initiatives ought to be consistent with the goal of collaboration. With this purpose in mind, many companies are beginning to move toward cross-functional evaluations of NPI participants. A smaller step in the same direction is to include among the criteria for personnel appraisals items such as "getting the job done under adverse conditions," or "fixing a problem." In less formal structures, when the priority of an NPI effort is made clear by top management, some employees will choose to volunteer to help in the design of a new product even though it is not in their job descriptions.

Finally, some incentives are already in place as intrinsic professional rewards: design engineers who are associated with a winning product in a dynamic industry have the opportunity to work on the next product and the newest technology; and manufacturing engineers, in similar circumstances, feel good about having contributed to doing things better and being a "key player" in the company. Also at work in some companies is: the satisfaction of "a job well done"; the desire to keep one's job in a company that will have to downsize if the new product introduction is not a success; and increase in influence (of an individual or a group) in future NPIs ("get to wear a bigger hat").

SUMMARY

The cross-functional team is a common organizational vehicle for translating product development targets into action. Such teams offer opportunities to reduce delays in information processing, improve problem solving, and allow more work to proceed simultaneously. In large organizations, the members of a core team may spend considerable time getting information and assistance from others whose time is not dedicated to that single NPI effort. Ultimately, however, it is the work of the cross-functional team that will spell success or failure in meeting the project's various objectives. Without a smoothly functioning, collaborative team, an overall goal of product integrity will prove to be elusive.

It is not enough to say that a carefully selected and empowered team, headed by a strong and talented leader, and linked to a supportive culture will be the means to introduce new products effectively. The use of cross-functional teams does not automatically produce the collaborative action needed to meet the targets.

Many NPI efforts are launched as a deliberate response to external pressure from a corporate parent or a major setback in the marketplace. Such crisis conditions can force collaboration on an NPI team and impose strong incentives not to fail. It is clearly more desirable to achieve a conducive new product culture without waiting for the external situation to become stormy.

It can be difficult to build collaborative behavior in established, functionally organized, multisite organizations. A senior manager in one world-class company began a speech reviewing a recent new product introduction by saying, "We started with a really dirty sheet of paper." Understanding and coping with the organizational, cultural, cognitive, and behavioral factors that this manager had in mind is important for all who are involved with product design and development. Promoting cross-functional collaboration also requires a productive balance between firmness and flexibility in managing NPI efforts. All key participants must have strong personal incentives to behave in ways that maximize the chances of an NPI success.

Once a collaborative workplace culture is developed, managers and team members can concentrate on substantive matters of design and development. In doing so, issues in the selection and implementation of technology are paramount. This is the subject of the next chapter.

CHAPTER 5

COPING WITH TECHNOLOGY CHOICE AND RISK

Managing technology is fundamental to the introduction of new products. For products as varied as home appliances, consumer electronics, motor vehicles, or specialty chemicals, much of the core day-to-day work of product design and development involves the identification and solution of technological issues. Some of these issues will center on the product itself; that is, its materials, components and their integration, size, and other constraining factors. Others will deal with the process through which they will be produced, that is, the extent and type of automation, mechanisms for high-volume testing, and modifications to existing equipment. In continuous process industries, such as chemicals and pharmaceuticals, technological issues of a product and its production process naturally blend together, while for products that are fabricated or assembled a special effort must be made to deal with them concurrently.

In all such contexts, there is a need to cope with technology choice and the element of risk as it may affect the critical success factors of an NPI project. *Technology* is broadly defined here to include the know-how, designs, hardware and software associated with the materials and components, end products, manufacturing processes, and information systems involved in the design and manufacture of products.

The existence of technological risk is most apparent for those products and production processes that are becoming increasingly complex. Yet even for rather simple products and processes, problems in achieving the time, cost, and quality targets for their introduction can often be traced to aspects of technology management. A failure to implement technology as planned will, at minimum, delay a new product introduction. It may also result in an increase in the cost of the NPI project. If the problem is not solved, product quality may suffer in ways that matter to the customer, and if production yields are affected, unit costs will suffer. Even if the problem is solved, the solution may be more costly than the planned approach and consequently, the unit cost may turn out to be higher than its target level. Customer satisfaction with new products often hinges on selecting the appropriate technologies and implementing them so they work as intended.

Chapter 2 addressed questions of *when* management should try to resolve *what* kinds of issues in product design and development. Chapters 3 and 4 then explored *how* project management and problem-solving teams can become effective vehicles for resolving specific design questions. This chapter adds the key variable of technology to our conceptual framework, builds on material presented in Chapters 2, 3, and 4, and addresses the following two questions:

1. What are the issues in managing technology for inclusion in a new product and in the production process through which it will be made?
2. How should these issues be addressed?

TECHNOLOGY MANAGEMENT IN NPI PROJECTS

Figure 5–1 schematically indicates the subset of technology management activity that relates to the definition and execution of NPI projects. Because our focus is on the single NPI project, we need not dwell on many of the topics that have traditionally been central to the literature on technology management. Two differences are worth noting at the outset to clarify our particular perspective.

First, the single product is not the best strategic focus for technology development (Maidique and Zirger, 1985). For many kinds of products, the most critical decisions about new technology occur when planning the next product family, platform, or architecture. Nevertheless, in the context of each individual NPI project, significant issues of technological choice and risk will typically arise.

Second, we do not address issues of communication and performance in the management of research, as distinguished from product development, settings (e.g., Katz and Tushman, 1979, or Tushman, 1978). The functioning of research laboratories and other ongoing programs of advanced technology development are clearly beyond the scope of this book.

Our framework for technology management in the context of new product introduction consists of actions in three domains (1) identification

FIGURE 5–1
Boundaries of Technology Management for NPI

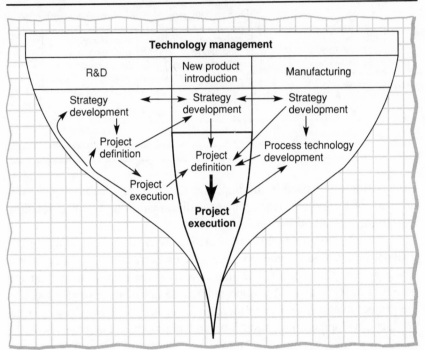

of technology requirements for a new product; (2) formulation of technology development plan for a new product; and (3) implementing the technology for a new product (see Figure 5–2).[1]

These domains have an implied sequential relationship starting with the initial consideration of technology requirements, proceeding to a development plan, and then to its implementation. Feedback among these domains of technology management is to be expected. For example, in formulating a technology development plan for a new product, participants may agree to modify one or more of the original technology requirements. The development plan, in turn, may need adjustment in light of

---

[1]The framework shown in Figure 5–2 and the findings relating to it originally appeared in a paper by Douglas Boike and Stephen R. Rosenthal (1990).

FIGURE 5–2
A Dynamic Model of Technology Management for NPI

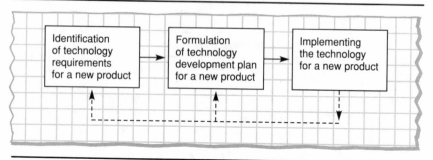

issues or options raised in the implementation stage. Each of these three domains of technology management is discussed in terms of its relationship to the overall NPI process.

Taken as a whole, this set of activities encompasses much of the goal-setting, planning, and problem-solving activities in the design and development of new products which embody considerable advanced technology. For this reason, it is important that senior management, as well as the project manager and team members, become comfortable seeing the product development process through a technology management lens.

IDENTIFYING THE TECHNOLOGY REQUIREMENTS

Successful new product introduction is market-driven, while being sensitive to the realities of technological progress. This means that the needs of customers should shape the concept of a new product, but the technological capabilities already available will largely determine the product solutions that are attempted. Accordingly, the identification of technology requirements for a new product necessarily involves the creation of consensus from a blend of disparate perceptions: (1) market needs and timing, (2) competitive offerings, and (3) technological capability and strategy. Technology requirements need to be clearly identified by the completion of the concept development phase of new product introduction, and many companies begin this effort before the project is formally launched (Phase 0).

Companies vary in terms of how systematic they are in their articulation of technology requirements and upon the relative emphasis they place on this early stage of the NPI process, and on the resources that are allocated. The rigor with which product planners attempt to translate statements of market need to associated technological implications also makes a significant difference at this point. To be successful, companies need to invest time, money, and resources at the onset of a new product effort. As suggested in Chapter 2, this is often a difficult trade-off in that impatience will quickly develop, focusing on doing rather than planning.

When the product is to be used "hands-on" by the customer, the translation of marketing perceptions of customer needs into technology requirements can be subtle and fraught with risk. At the core of this difficulty are the different "thought worlds" between these two functional specialties, a problem that was discussed in Chapter 4. In a leading-edge company seeking first-mover advantage, R&D personnel may have the power to argue either that "small changes" in the technology requirements will be unnoticed by the customer (on the downside) or that they will greatly enhance the value of the product (on the upside). In the former instance, R&D may want to be conservative and adapt a proven technology solution rather than stretch to develop a new solution, given the time and resource constraints on the NPI project. Conversely, in the latter instance, R&D may be tempted to try to embody a new advanced technology in the upcoming new product, and will call upon a marketing rationale for so doing. Either way, the issue is whether the R&D personnel are familiar enough with the market and the customer use and reaction to existing products to be able to make a reasonable judgment. Marketing people who are involved in this decision should be conversant enough with the technologies in question to be able to raise appropriate questions of costs and risks.

Sources of Technology Requirements

A varying mix of demand versus supply arguments provide the foundation for different companies to specify the technology requirements for a new product. Companies that are traditionally market-driven will naturally tend to derive a sense of technology requirements from statements of what the customer would expect from a "quality" product. Since customers may care about a number of different attributes of a product under the umbrella label of quality (see Chapter 3), this full specification of requirements can be a complex task unto itself. Various types of market research, as outlined in Chapter 12, probably will be called into play at this time.

In contrast, engineering-driven companies are dominated by the pursuit of new technological capabilities. These companies will tend to think of technology requirements directly in terms of achieving certain improvements, using standard quantitative measures of product performance, relative to either their own or competitors' products. While such engineering-specified requirements will always address the technologies that are embodied in the new product, they will sometimes also include attributes of the associated process technologies for manufacturing or testing the new product. The inclusion of process technology requirements is particularly likely when the company views its manufacturing capability as a competitive weapon.

The determination of a set of technology requirements for a new product is a serious effort and requires preliminary consideration of the implications of alternative requirements upon the feasibility of the project. Remember, at the concept development stage, targets are set for the cycle time, development cost, and the unit cost of the new product. Technology requirements will exert direct impacts on all of these targets, although the connections may not be very clear until the development plan is specified.

Overstating the technology requirements for a new product is a trap to be avoided. Careful managerial attention at the early phases (Gates 0 and 1, using the terminology of Chapter 2) can make a big difference. Care must be taken not to "stack the deck" against the emerging NPI effort by starting with unnecessarily overly ambitious technology "requirements." Conversely, there is a danger of creating a "me-too" product, which fails in the marketplace due to its conservative stance on technological selection. In the situations covered in this book, where product design and development efforts involve significant product or process change, there seems to be a tendency to overreach, rather than underreach, in specifying technology requirements. This is especially probable when marketing tends to want the new product to be exciting and engineering is more optimistic than realistic with respect to the associated time, cost, and risk. Companies with a clear sense of direction regarding new technology usually find a pattern of incremental innovation to be faster and cheaper, over time, than a strategy of less frequent, major technological breakthroughs.

The Strategic Use of S-Curves

"Figures of merit" (FOMs) describe what is important about a product. Identifying them begins with an understanding of characteristics of a product, valued by a customer, that will drive technological innovation. Figure 5–3 lists a few such examples.

Note that different products will measure a FOM (e.g., operating speed, precision, and portability) in different ways, and each will have its own rate of progress over time as a result of underlying technological innovation. Historical analysis of the rate of progress of many technologies—using a relevant quantitative measure of a significant figure of merit—has shown a consistent pattern commonly known as the *S-curve*. Here we provide only a summary of the S-curve technique and its relevance for those engaged in the design and development of new products.

Progress in the achievable level of any figure of merit for a technology is the result of cumulative effort over time by many different organizations. In the early period of developing a new technology, the progress might result from research projects at university or corporate laboratories that made an early commitment to explore some technology that seemed to offer great promise. Progress is slow for a while because not much is known about the nature of the technology, and the learning that takes place in this early stage is relatively costly in terms of demonstrated progress (in moving to new levels of that figure of merit). The technology is at the first stage of the S-curve, as shown in Figure 5–4.

In the middle stage, one or more dominant approaches emerge and there is explosive productivity in generating technological progress. Competition often can be fierce during this second stage of the S-curve, as many different organizations try to produce technological advances by building on prior approaches that, in retrospect, appear relatively primitive. In some industries, a technology leader emerges and controls the pace of improvement, but the exponential growth in productivity of the technology development effort will still be apparent.

Ultimately, the technology is "played out" in the sense that the figure of merit of interest can no longer be improved as readily as before. In this third stage of the S-curve, all of the easy improvements have been made. Squeezing more improvement out of the same underlying technology, if possible, calls for much more intensive levels of resources. Alternatively, natural ceilings are reached in the capability of the underlying material, concept, or process.

FIGURE 5–3
Grounds for Figures of Merit (FOM)*

Sample Products Characteristics:	Power Lawnmower	Vacuum Cleaner	Refrigerator	Milling Machine	Chain Saw	Personal Computer	Camera
Operating speed				✓		✓	✓
Storage capacity			✓			✓	✓
Degree of precision				✓			
Reliability						✓	✓
Energy consumption			✓				
Portability		✓			✓		
Durability	✓			✓	✓		
Serviceability	✓				✓		
Ease-of-use		✓		✓			✓
Safety	✓						

*Note: Cost is often factored into an FOM.

FIGURE 5–4
S-Curve of Technological Progress

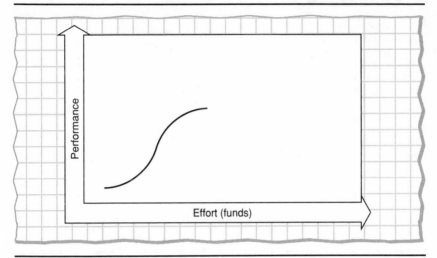

But technological progress is not necessarily over at this point. Switching to a fundamentally new material, concept, or process in the search for greater progress is the equivalent to starting at the beginning of another S-curve. This occurrence, often called a *technological discontinuity*, is shown in Figure 5–5.

This new S-curve will follow the same sequence of the stages, if indeed it shows promise. A farsighted laboratory or company might start work on a new S-curve before the prior S-curve has fully matured in the belief that it represents the wave of the future and the risk of being a pioneer is worth the potential benefit of being a technology leader. When two such S-curves cross, the emerging technology has matched the figure of merit of the former base technology and, if the promise is realized, will overtake the former technology eventually rendering it obsolete, all other things being equal.

Viewing technological progress from this historical analytical point of view, one sees the dilemma of the company that has committed itself to producing technological progress along the presently dominant S-curve. How long should it continue to base its products on this dominant technology and try to make incremental improvements to that technology? Is it more sensible for a company to begin to work on technological progress

FIGURE 5–5
Technological Discontinuity

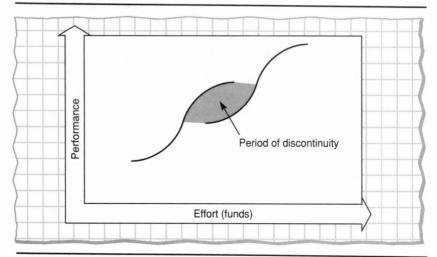

along a newer S-curve and leave the older, more mature technology behind? This kind of thinking is what we refer to as the strategic use of S-curves.

This strategic use of S-curves has been discussed at length by Richard Foster (1986). Essentially, the S-curve is an empirical tool for helping senior management understand the risks and benefits of being an "attacker" by pursuing new, high-potential technology, as distinguished from one who strives to "defend" the advantage already gained through the development of an existing technology. Foster is fundamentally concerned with why leading companies abruptly lose their markets to new competitors. His conclusion is that they cannot resist the trap of remaining a "defender," in the erroneous belief that this is always the safest path to follow.

In summary, the S-curve dynamics are lurking in the background as a team begins to identify the technology requirements for a new product. If the company is skilled in technology forecasting and has already done extensive S-curve analysis, the team can take its findings into consideration as they move toward a technology development plan for *that* NPI project. If this technique is not being used, the company should consider including it in an appropriate training program. Admittedly, however, it is much easier to conduct S-curve analysis looking backward in time rather than ahead, in the forecasting mode.

If no such analysis exists, senior management should, at minimum, be sure that the related questions of technological risk and discontinuity are raised early in the NPI process. Either way, the team can be guided to include important issues of competitive dynamics in their formulation of technology requirements and associated plans for product design and development. To the extent that there is a clear identification of technology requirements during concept development, the development phase will benefit from a more coherent vision and focus. To the extent that this is not accomplished, subsequent iteration and delay commonly occur.

Other Relevant Kinds of Analysis

Although a variety of other analytic approaches can be used to support the identification of technology requirements for a new product, it is beyond the scope of this book to cover any of these in detail. A few basic considerations, however, are appropriate to mention at this point.

Demand-side analysis looks to the market as a basis for anticipating emerging technology requirements. These methods include: market/customer research, performance/cost modeling, and specification comparisons. Customer surveys and users' groups can provide direct information on what additional product capabilities or other characteristics are desirable. It is also important, however, to observe current customers and conduct research with potential customers to understand what their preferences are. Lifestyle analysis and perceived value analysis are two useful approaches for identifying new technology requirements. For industrial products, the importance of consulting lead users has been well documented (Urban and Von Hippel, 1988). Since surveys and broad market research have their methodological limitations, many companies use focus groups: a sample of desired customers are assembled for several hours to serve as a laboratory experiment aimed at eliciting subjective responses to alternative capabilities and styles of a potential new product. These customer-based techniques for identifying technology requirements are often inappropriate when radically new products are being considered.

The customer usually expects to pay more for products with enhanced performance but the feasibility of achieving certain cost targets when enhancing the product's performance needs to be checked. The ultimate check on this comes during pilot production, before volume manufacturing begins, but this is much too late to be of strategic value. Accordingly, the development of analytic models for predicting performance/cost relationships early (at the conceptual design phase) is becoming more common in many industries.

Finally, companies need to adopt a competitive perspective on the question of technology requirements. One way to do this is to make explicit comparisons of the performance specifications of all existing and announced products in the class of interest. Such comparisons, coupled with data on market acceptance, can guide the identification of technology requirements for the new product.

Supply-side analysis looks at patterns of technological innovation and recent technological breakthroughs as a basis for setting requirements for a new product. These methods include: (a) assessing the availability of key enabling technology; (b) projecting the risk of new applications of existing technology; and (c) examining the "fit" of technology applications with the company's long-term research and development strategy. The availability of key-enabling technology can be determined by drawing upon R&D experts within the company, within supplier organizations, or specialists who are consultants in this field. Risk assessment of new technology applications is currently more of an art than a science, yet it is an essential element of the conceptual design phase. Even before a new product introduction is formally begun (i.e., in Phase 0) supply-side analysis should include explicit matching of potential areas of technology "stretch" with the company's long-term strategies for R&D and strategic alliances.

Market-Pull versus Technology-Push Requirements

As suggested by the above discussion of demand-side and supply-side approaches, a company needs to adopt a set of analytic approaches that suits its own market and competitive context. The identification of technology requirements generally follows one of two philosophies: (1) competitive success is dominated by the varied needs of a broad customer base, or (2) it is determined more directly by the performance specifications of the product. These two approaches to the identification of requirements are normally referred to as *market-pull* or *technology-push*.

Under a purely market-pull scenario, identification of technology requirements is a two-stage process, starting with an assessment of customer needs and then translating those needs into technological terms. The identification of customer needs (using techniques of the sort outlined previously) is performed by marketing or product planning staff. Then, armed with data on customer needs, a product manager will bring technologists together with marketing people or product planners and convert the statements of need into technological dimensions. Approaches such as Pugh Analysis or quality function deployment (QFD), discussed in the appendix to Chapter 6, have been developed to aid this specification process.

In contrast, under a purely technology-push scenario, clear performance metrics have already been defined in the industry and the identification of technology requirements essentially involves setting appropriate performance targets using these accepted metrics. Product engineers will adopt some or all of the following techniques as a basis for specifying technology requirements of this type: competitive teardowns, benchmark comparisons and cost/performance projections (including the use of S-curve forecasting techniques).

Companies often refer to their corporate culture as being either market-pull or technology-push, but this should not determine how technology requirements are set for a new product. R&D (research and development) and marketing personnel need to collaborate with each other, especially in the critical early stage of specifying technology requirements. Common misunderstandings between these personnel often lead to various "states of disharmony" that adversely affect project success rates (Souder, 1987).

To be effective in identifying the technology requirements, a company's approach must be compatible with its industry and the nature of their product, rather than the functional background of its senior management. One basis for approaching requirement specifications is to ask whether the product is used directly by a person as an end product or as a component of an encompassing product or system (and therefore used only indirectly by the ultimate customer). Industrial products tend to fall in the latter category, whereas consumer products tend to fall in the former category, but this is not always the case (see Figure 5-6).

When a product is designed to be used directly by a person (i.e., consumer, operator, and so on) as part of a man-machine system, figures of merit include dimensions other than performance and cost. Examples include: ease-of-use, features/functionality, styling, and dependability (in contrast to pure engineering reliability). The success of such products will

FIGURE 5–6
Establishing Technology Requirements for a New Product

		Type of Product	
		End Product (Direct Use)	Component of a Larger "System" Product
Complexity of the Product	**High** *Technology specification is more difficult*	Translate customer preferences • Business telephone system • Material conveyor	Translate system requirement into component requirements • Jet engine • Mainframe computer
	Low *Technology specification is easy*	Use customer preferences* • Athletic shoes • Breakfast cereals	Use customer's specification of component requirements† • Stereo speaker • Car battery

*For end product, translation is easy and direct.
†For system components, customer provides technology requirements.

depend upon delicate, and often difficult to quantify, trade-offs of user-perceived value/benefit with the intricacies of the needed technologies.

Effective reconciliation of such trade-offs requires a more cross-functional process of technology requirements identification. Sometimes it can be difficult to establish the design alternative that will be most favorably received by the customer. As discussed in Chapter 4, the application of ergonomic principles and human factors approaches can suggest enhancements in ease-of-use. In the early stages of designing products with a high level of human interaction, industrial designers can suggest shapes and materials, and field service organizations can provide feedback on problems that customers experienced with prior products. All of these considerations need to be translated into particular technology issues and options. The management of this process of technology identification can be very demanding, and the project management skills identified in Chapter 3 will be of value.

Now let us consider a different kind of situation, where the particular product to be introduced will be embedded in a larger "system" product and is therefore used in a more indirect sense by the customer. Here success of the new product depends on how it fits within the broader product or system, that is, on its performance attributes. In cases where such a product is being designed to meet the needs of a specific customer (e.g., a jet engine to power a particular aircraft), the performance specifications will be provided by that customer.

In other cases, the product is aimed at a broader market or users (e.g., a high-end computer) and the performance specifications need to be established by the developer. Primary factors to be considered here are the developer's own technology strategy (leader or close follower) and the probable state-of-technology at the time the new product will be released (determined, perhaps, through S-curve projections). Here, one of the early product design challenges is for marketing and engineering representatives to reach a consensus, through negotiation if necessary, on the performance specifications for the new product.

Another concurrent basis for defining the technology requirements for a new product is the inherent complexity of the product itself. The important distinction with respect to product complexity is whether the customer can directly identify the technology requirement (low complexity) or only the effect of that technology (high complexity). Figure 5–6 combines the two dimensions of use-of-product with complexity of product to show the various approaches to identifying technology requirements for a new product.

In any event, identification of technology requirements is a step early in the conceptual development (Phase 0 or 1) of a new product. Figures of merit for such products are dominated by a measure of some aspect of performance (e.g., an aircraft engine's fuel consumption or end-of-climb thrust) or cost-performance ratios, such as dollars per million instructions processed ($/MIP) for a large-scale computer.

In summary, the identification of technology requirements for a new product can employ a number of available techniques, and a fundamentally different approach is followed when the product is to be used directly by a person versus when it is not. Furthermore, when some of the requirements for a new product are expressed in terms of its serviceability or ease-of-use, a different sort of "technology requirement" needs to be formulated. Rather than being a direct specification of a quantitative figure of merit that the new product must achieve, the ease-of-use of serviceability criteria (discussed elsewhere) serve as indirect constraints in subsequently selecting one technological solution or another.

The desired outcome of this first domain of technology management

FIGURE 5–7
Interrelationships among Major NPI Decisions in Planning Phases

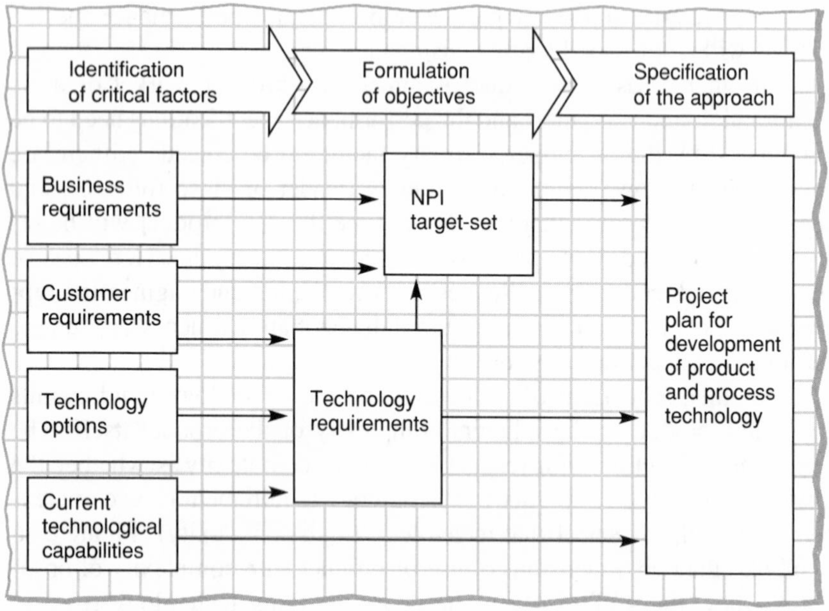

is a customer-driven set of technology requirements, using accepted FOMs, to drive product design and development (see Figure 5–7). Whenever technology availability is a concern, early involvement of technology experts can help to reconcile issues of needs versus capabilities. This sometimes formal, sometimes informal, process can drastically reduce the subsequent iteration time to reconcile difficult or unachievable technological objectives. At the center of such considerations should be the notion of technology that will be perceived by the customer as value-added.

FORMULATING THE TECHNOLOGY DEVELOPMENT PLAN

Once the technology requirements have been specified, attention shifts from an emphasis on the "what" of specifying requirements to the "how" of NPI project planning. Starting in the concept development phase of the project and continuing into product and process development, technology options are identified and evaluated. Decisions and commitments are then made to pursue particular product and process technology "solutions." Arguments and analysis bearing on economics, levels of effort, internal and external resources, time requirements, other NPI priorities, technology capabilities and strategies of the firm, projected competitive responses, and aspects of risk and uncertainty should all be invoked at this time.

Companies vary in terms of how thoroughly they specify and assess their technology development options. Existence of one or more of the following factors, discussed in earlier chapters, tends to *encourage* broad systematic inquiry:

1. A rigorous formal NPI review process tied to the release of resources.
2. A workplace culture that emphasizes cross-functional problem solving.
3. A generous allocation of experienced professionals with the requisite technical skills.
4. A complex product requiring considerable technology development activity.
5. Ongoing efforts to involve the customer in this phase of planning.
6. Consideration of technological options offered by alliances with suppliers.
7. A project manager able to serve as a catalyst for identifying and resolving trade-offs in technological approaches.

Factors that would *impede* broad systematic inquiry in establishing technology development plans include:

1. Relentless pressure to get development activity started due to an impending product launch date.
2. Budgetary pressure to spend money for development activity rather than for planning.
3. Prior commitment to certain product or process technology initiatives.
4. The nonexistence of any of the factors of encouragement listed above.

The Context for Technology Choice

The time available for planning the NPI project is bounded but participants should be careful to consider whether the targets for development cycle time and cost and the unit product cost seem compatible with the technology options that are selected. The specification and assessment of technology options, therefore, is inherently context-bound and opportunistic. Some companies help ensure that such compatibility will be obtained by arbitrarily restricting any new product to using no more than, say, three technologies not already tested in commercial applications. Clearly, such a rule of thumb is no more than a rough proxy measure for the inherent risk and complexity being absorbed into any single project.

Furthermore, information collected and conclusions reached when formulating a technology development plan may lead a company to restate its NPI technology requirements to be consistent with that plan. Ideally, a company's existing technology strategy will have already stimulated internal R&D programs and key external alliances that can be drawn upon to support the development of new products.

The formulation of a technology plan for a new product should occur early in the NPI project. The group working in Phase 0 to validate the new product idea needs to begin the specification of the technology plan. Then, in Phase 1, the core team ought to refine this plan after considering the trade-offs among the various goals of the NPI. Each functional requirement of the product will yield options for both product and process technologies. In many companies the research organization routinely maintains a data base of new technologies that might be appropriate for future products. By including all appropriate specialists in the formulation of the technology plan, such choices among options can reflect factors of interest to the customer, such as development cost, unit cost, product life cycle cost, time to market, reliability, or ease-of-use.

Frequently, the team will realize that some of the information relevant to these choices is not available at this early phase of the project. Under such circumstances, there is no choice but to proceed to product design and development with aspects of the technology development plan being incomplete. Consequently, portions of the technology plan will be explicit from the start and others will be added, based on the subsequent specification and analysis of available options. When uncertainty associated with technologies is high at the beginning, project planners should identify safer options as contingencies that might have to be used if the preferred technology option turns out to have formidable problems. In any event, careful and thorough technology planning will facilitate project execution and reduce the development cycle time.

Strategic Considerations

Early decision will shape a company's technology development plan for a new product. Sometimes the early decision will be to select a core technology in which the company is already established or wishes to establish a strategic advantage. Such strategic thinking is closely aligned with S-curve analysis. It argues that the best use of incremental development efforts is either to try to push the firm further along the S-curve it has already committed itself to or, conversely, to shift attention to a newer and more promising S-curve. In Chapter 8, we will see that Northern Telecom Inc., already a leader in the development of digital switch technology for large-scale telecommunication systems, decided to incorporate digital technology (with new architecture) in its new small business system where analog technology has formerly been prevalent. Given a different competitive and market scenario, Amdahl Corp. planned its 5890 computer system to be an enhancement (with new architecture) of its existing 580 computer, rather than a new product that would contain any radically new technology (Venugopal, 1990). The key for all companies is to recognize those technologies in which a "true" strategic advantage exists in either cost, functionality, reliability, or design.

Sources of Technological Risk

When there is severe pressure on development cycle time, an emphasis on risk reduction is often advisable. Here, one ought to consider an early decision that is more conservative than the above; namely, to plan to use designs that have been successful in prior products. Naturally, this kind of early decision must be mindful of an opposite risk: the underachievement of performance and cost targets. When General Electric (GE) was planning the design of its CF6-80A jet engine, as described in Chapter 10, they decided to work from a successful prior engine and shorten the length and eliminate the turbine midframe. In another industry, Motorola Corp., a leader in the design and manufacture of pocket pagers started with an existing receiver design as a basis for a new product line (see Chapter 9). Northern Telecom, in the example above, decided to design their new small business telephone system so that it could use most of the same production process that was in place for the prior product.

Another rule of thumb is for the NPI team to identify, as early as possible, those product components or process capabilities that will require new development of technology. Special risks to be incurred and rewards to be gained from such an approach need to be made explicit. So doing seems to be less common than reason would suggest. Depending on a company's situation with respect to technology, different questions need to be asked at this time. For example:

What is involved in adopting a new kind of component for our product? When Agfa Compugraphic, a leading maker of desktop publishing systems decided in 1988 to incorporate a laser diode for the first time in one of its products, they had to decide in advance how difficult this might be. It turned out that the product designers, through lack of experience with optics, had rather optimistic notions about the ease with which they could take this step into technology which was, for them, new (Tatikonda and Rosenthal, 1990).

What is the risk of using new materials in different ways in our product design? When the designer of an advanced commercial aircraft engine designed the fan blades to be made of layers of titanium and synthetic materials (rather than a single solid cast) the risk of loss of strength had to be assessed. Despite the early assessments made by the company, some customers remained skeptical that this choice would prove to be a good one.

What is involved in solving certain problems with our manufacturing process? Successfully dealing with this aspect of technology planning becomes easier when a company is familiar with the material or component in question. A manufacturer of industrial shelving, for example, will probably have the in-house expertise to estimate the effort required to design and manufacture a new product to overcome prior problems of misalignment in an existing product of the same material. However, when considering the installation of production processes that are new to the plant, it is important to carefully consider the likely risk and potential benefits. Avoiding tendencies toward undue optimism or pessimism is a special challenge here. When Northern Telecom was designing its Norstar product (see Chapter 8), they decided to introduce a moderate amount of surface mount technology at the plant, but only after rather extensive negotiations between design engineers and manufacturing engineers.

What is the risk associated with relying on external suppliers for needed technology? Planning from the beginning to outsource those technologies where no internal sourcing advantage exists is a common early decision rule. But this does not absolve the team from a careful consideration of technology choice and risk in the planning stage. The NeXT Corp. intentionally took a certain amount of risk in procuring advanced components from two of its key suppliers. However, they carefully selected those suppliers, Canon and Fujitsu, for their pioneering capabilities in those technologies and they were convinced at the time that the choice of these components would be important in meeting customer requirements (see Chapter 7).

Principles of Technology Development in NPI

The previous discussion suggests certain principles be followed in developing a technology plan in support of an NPI project:

- Leverage existing base technology already on hand wherever possible.
- Look for new technological options to support unique product requirements.
- Control and challenge those technology areas that have maximum strategic advantage without pushing the program viability and achievability over the limit.
- Use technology easily available from external sources wherever appropriate, for example, in areas that are noncritical and/or areas where a significant reach is needed and appears unachievable.
- Take special precautions to moderate risks taken with technologies that are secondary to the success of the project.
- Manage ongoing technology efforts (internal and external partners) to maintain a high level of awareness of technology risks and status.
- Incorporate experience from prior products in assessing technology risks, and when the risks appear high, include contingency plans from the beginning.
- Maintain a budget and management system to control technology activity in new product development projects.
- Seek to develop technology that can be modified in modular form

thus enabling future additions or subtractions that would enhance product functionality, cost, and performance.

* Consider the alternatives for incremental technological change where the initial product offers less than what customers currently require.

Implications for Ongoing Advanced Technology Development

Technology planning decisions should be taken in the context of longer range "road maps" or strategies. Accordingly, while pursuing these principles in an individual new product introduction, companies also need to develop a plan for sustaining their internal portfolio of technological competencies. The portfolio must be focused with due care in areas that have maximum strategic impact. This may be facilitated by a periodic planning exercise to develop and update technology and product maps (Wheelwright and Sasser, 1989).

Strategic alliances in such critical areas of technology development are increasingly common as product manufacturers choose to rely upon external expertise for leading-edge knowledge of advanced component processes or materials. Companies such as Corning have made technology alliances the cornerstone of their strategy for forging a leadership role in a number of different areas while sharing the cost and risk with long-term partners.

However it may be implemented, a technology strategy should address likely needs over an extended period of time, including potential future product programs that are not yet on the drawing board. It should also reflect learning from current or past product development programs and therefore be a dynamic rather than static planning process (Adler, Riggs, and Wheelwright, 1989). By having such an evolving strategic portfolio of technologies, the level of risk on any particular NPI project can be considerably decreased. Technology strategy formulation is beyond the scope of this book and we have touched on this subject here only to highlight an important interface between product design and development and a related topic.

IMPLEMENTING THE TECHNOLOGIES: LESSONS LEARNED

This is the action domain in which technologies are designed, developed, and tested within the NPI project. As discussed in Chapter 3, the unit cost and quality of the resulting product, and the ultimate cost and time-to-market become measures of the success of technology development activities. It is here in project implementation that one would expect to reap the rewards from the careful specification of technology requirements and development plans.

The ongoing attention of the project manager is especially important when the new product being developed calls for the use of an immature or loosely characterized technology. Special effort should be made to identify and respond to emerging problems of technology development as soon as possible in the project. Success in this pursuit will be facilitated by a number of factors which have already been discussed:

An extensive degree of simultaneous engineering between product and process development.

An emphasis on early and rapid prototyping.

Avoiding new conflicting NPI priorities from elsewhere in the business unit.

Establishing and sustaining an appropriate priority for this project within the involved manufacturing organizations.

Continuity of key technical staff from earlier phases of the project.

Another ingredient of success, to be discussed in Chapter 6, is the use of CAD/CAM (computer-aided design and computer-aided manufacturing, respectively) and other information-based design technologies.

Improving technology management in new product introduction often requires more attention to matters of personnel allocation and team formation. General issues in selecting the project manager and team members were discussed in Chapter 3 (relating to the project manager) and Chapter 4 (with respect to the project team). Other more specific lessons are:

Project participants may require education in advanced technologies to prepare them to better handle their assignments.

The project manager must be prepared to secure the unplanned services of a technical expert, when the existing team lacks sufficient experience to resolve a design controversy.

Manufacturing launch should not proceed until there is evidence that the various technologies are under control. This is equivalent to saying that the design project has to be constructed to get accurate early readings on the technology readiness of both the product and the manufacturing process. Discovering such problems during or after product launch is a disaster to be avoided even if the upfront cost to do so is considerable.

Supplier technology must be considered an essential element of the portfolio. This has implications for the inclusion of key suppliers early in the NPI process as well as procedures that ensure tight controls on the quality of supplier-delivered technology.

The management of manufacturing process technology must, in many instances, be considered coequal with product technology. (This requires special attention to testing the process capabilities well before scheduled product launch. Process prototypes may be as critical as product prototypes.)

Success in implementing technologies involves the management of knowledge as a critical resource. With access to needed knowledge, the complexity and uncertainty inherently part of most NPI efforts can be kept under control. Without such access, technological issues embedded in the design and development of a new product can thwart genuine efforts to meet the targets set for such a project.

SUMMARY

Many aspects of technology strategy—such as core competencies, investment in long-term development of product and process capabilities, sourcing internally versus through partners—are decided by senior management. Yet there are still many important decisions to be made at the project level, when products are being designed and developed. At this level, technology management consists of making choices among technological options and managing the associated risk for a particular product. Some of the questions that arise are:

Choices. Do we scan technology options broadly at the beginning? How do we go about narrowing the choices? Do we have good screens that are activated at the right times? What are these choices based upon? Is our process systematic enough? Do we need to revisit our choices more or less often? Are we setting a technology platform?

Risks. Do we understand the technological risks associated with our decision making? Are we overreaching or underreaching technically? What contingency plans do we have for high-risk, critical items? What corporate approaches to technology development facilitate smooth project implementation?

These questions essentially ask how the decision-making process can be improved to incorporate appropriate technology more effectively, while minimizing unpleasant and untimely surprises. Although there is no single answer to these questions—other than the frustrating "It depends"—this chapter has presented a simple framework for looking at technology choices and risk. This framework integrates earlier discussions about project phases and managing to targets. In short, one cannot eliminate all risk from product design and development. Yet one can take certain steps to identify, moderate, and manage the inherent risk through certain key steps in Phases 1, 2, and 3 of the project.

Clear technology requirements, stating what the product must be able to do, are the foundation for downstream vision and focus. The needs and expectations of customers should be reconciled with technological possibilities and capabilities to establish clear technology requirements early in the project's life. Products that are embedded in a broader system can often cast their requirements directly into engineering terms. Usually, however, it is desirable to blend market needs and timing, competitive offerings, and technology strategy and capabilities into the requirements. This should be attempted first in Phase 0 and then refined in Phase 1.

After a range of product and process technology options have been identified, they are assessed against the targets for the project, internal and partner capabilities, and abilities to manage risks. This requires translating the trade-offs that are already embedded in the targets into a particular physical reality. In other words, how will various technological choices impact development cost, unit cost, product life cycle cost, time-to-market, reliability, and ease-of-use? The ideal technology development plan represents the "solution" to balancing all of these factors. The technology plan is, in this sense, where the rubber meets the road: How well does this set of choices reflect what we said we wanted to do?

Finally, the product and process technologies need to be designed, developed, and tested. Technology implementation is facilitated by flexibility in adapting to shifting project needs. As the project moves through its work phases, different skills need to be tapped, and moved into the foreground. This could mean shifting the leadership of the team, whether formally or informally, or using outside technical experts, or relying on supplier technology and expertise. Effective project management then becomes critical in retaining the voice of the customer, while guiding the project to a timely and productive completion. This pursuit is greatly aided by the use of design technologies and practices, which is the subject of the next chapter.

CHAPTER 6

SELECTING AND USING DESIGN TECHNOLOGIES*

All phases of the design and development of new products require that groups of functional specialists make a large number of decisions. Such decisions include the form, fit, and function of the product and how to manufacture it within cost, time, and quality targets. In making these decisions, participants conduct design and development activities (some quite routine, others not) that can be enhanced through the adoption of new technologies and practices.

The acquisition of advanced design technologies and practices is often claimed to be a solution to many of the issues raised in this book. While this may be true to some extent, management must treat this subject with a more critical eye. A more in-depth understanding of the purpose and pitfalls of design technologies is often necessary for its effective adoption and use. In other words, concepts of technology choice and risk should be applied to the *process* of product design, as distinguished from the *content* of the new product and its associated manufacturing system.

This chapter builds on earlier discussions (Chapters 2–5) of structuring the work of NPI, planning and executing the projects, using cross-functional teams, and managing technology. Here we look more closely at the information processing dimension of product design and development. With this perspective, we present a framework for understanding how the creation of competitive products is related to the design process that is employed.

We use the collective term *design technologies and practices* to be the approaches employed, either in a singular or an integrated fashion, to improve a product through its design and manufacture. These approaches can be a method, model, or even a computer software package used by those engaged in product design and development activities. Product im-

---

*The material in this chapter and the appendix was also published in a paper by Stephen R. Rosenthal and Mohan V. Tatikonda (1990).

provements include: lower unit cost, higher functionality, shorter manufacturing cycle time, shorter product development cycle time, lower development cost, and higher end product variety.

Companies seeking to build a coherent and growing capability for designing and developing products, must decide which of the many available design technologies and practices to adopt. But these design technologies cannot be independently ranked or rated in terms of their inherent value. Instead they must be assessed in terms of how they are used in actual projects. This chapter and its appendix address two questions:

What are some of the existing design technologies and practices?

What is a useful framework for appreciating their potential contributions to the design and development of products?

A FRAMEWORK FOR ASSESSING DESIGN TECHNOLOGIES

The selection, design, and development of new products is information intensive. It requires accumulation of data and insight from diverse sources and different functional perspectives. As a new product is being designed and developed, information is constantly accessed, interpreted, augmented, transformed, and deployed. The accuracy, consistency, and availability of such information essentially determines the extent to which a new product achieves required functional capabilities, ease of manufacture, and a fit with overall strategies. Adopting an information processing perspective on new product development can facilitate the strategic assessment of different design technologies.

A set of six functions is central to the successful design and development of new products. These functions, each signifying a different form of information processing in the product design and development process, are:

1. Translation.
2. Focused information assembly.
3. Communication acceleration.
4. Productivity enhancement.
5. Analytical enhancement.
6. Management support.

Although these six terms are not in widespread use, the underlying activities are familiar (at least in tacit form) in many companies. By paying more explicit attention to each of these functions and ways of achieving them, companies can develop an approach to pursuing their competitive strategy through the adoption of particular design technologies and practices. We will explain the significance of each of these functions and provide illustrations as to how they may be achieved.

Figure 6–1 summarizes the predominant relationships between an illustrative set of design technologies and practices and the six information processing functions. The citations in this table could be expanded to include links of lesser significance than the ones shown and discussed later. Furthermore, other design technologies and practices could be analyzed in a similar manner. In short, an important set of relationships is shown in Figure 6–1 and in the discussion that follows, but this should not be considered to be a fully exhaustive analysis.

Any significant design technology or practice, successfully implemented, can strengthen one or more of these information processing functions by using existing information to bring about more efficient or effective new product developments. Depending on its context of use, configuration, and form of implementation, a given technology/practice can support different information functions at different levels (as illustrated in the appendix to this chapter).

Naturally, the benefits from any particular design technology or practice will vary depending on its use. Broadly speaking, however, one can say that ultimately a technology or practice is important either because it leads to the development of the same product at a lower development cost or in shorter time frames (NPI efficiency) or that it contributes to the design and development of a better, cheaper, of different product (NPI effectiveness). Such outcomes can be perceived by the customer. They are centrally related to the objectives of new product development and we refer to them with respect to each of the information processing functions.

Translation

The product development process inherently requires that various kinds of specialized functional work be performed. A coherent project must assure that these specialized planning, design, and development activities

---

FIGURE 6–1

Design Technologies and Information Processing Functions

Translation
  Quality function deployment (QFD)
  Design for assembly (DFA)
  Customer use into test
    requirements
  Target costs into yield objectives
  Computer-aided process planning
    (CAPP)
  Planning bills of material (BOM)
  Value engineering

Focused information assembly
  Early vendor involvement
  Early manufacturing involvement
  Simultaneous engineering
  Co-location of design and
    manufacturing engineers
  Quality function deployment (QFD)
  Design for assembly (DFA)
  Design reviews
  Manufacturing systems simulation

Communication acceleration
  Computer-aided design (CAD)
  Group technology (GT)
  Electronic data interchange (EDI)
  Early specifications to vendors
  Computer-integrated manufacturing
    (CIM)
  Planning bills of material (BOM)
  Preliminary prototypes
  Rapid prototyping
  Early product information to field
    service
  Early product information to
    marketing/sales

Productivity enhancement
  Drafting computer-aided design
    (CAD)
  Computer-aided software
    engineering (CASE)
  Project evaluation review technique
    (PERT)
  Computer-aided engineering (CAE)
  Group technology (GT)
  Computer-aided manufacturing
    (CAM)

Analytical enhancement
  Manufacturing simulation
  Learning curve analysis
  Computer-aided design (CAD)
  Finite element analysis
  Robust engineering
  Statistical design of experiments
  Taguchi methods

Management support
  Gantt charts
  Project evaluation review technique
    (PERT)
  Contract books
  Formal performance reviews
  Milestone gate reviews
  Design for manufacture (DFM)
    checklists
  Manufacturing sign-offs
  Group sign-offs

---

are mutually consistent and compatible. This can be encouraged, particularly in a small company, by making sure that the various specialists talk to each other regularly. Even then, as described in Chapter 4, the project effort is likely to be hindered by the inherent breakdowns in communication brought about by the isolated "thought worlds" that typically dominate the problem-solving orientation of functional specialists. What is needed at such times are design technologies and practices that facilitate the transformation of sets of information from one point of view to another. This information processing function, which we call *translation*, should be explicit and as routinized as possible.

For example, marketing product planners aim to determine customer desires for new products. These desires are often communicated in general terms that have little to do with implementable product capabilities. Usage of quality function deployment techniques would facilitate translating; for example, "nice writing flow" for a pen to "xx viscosity of pen ink and yy roller ball pressure in pen." Subsequent to this marketing-to-product-design translation, a company might employ a design-for-assembly algorithm to translate product design specifications to particular manufacturing engineering requirements. Similarly, projected variations in usage environments for some new products need to be translated into specifications for various in-line test equipment. Target unit costs for a new product need to be translated into yield objectives for manufacturing ramp-up. Similarly, computer-aided process planning (CAPP) can translate product designs into manufacturing routings and detailed process plans. Planning bills of material (BOM) may translate preliminary product part hierarchy to purchasing requirements and assembly facility layout. Value engineering can translate product functionality and service requirements into product cost and materials guidelines.

As suggested by these examples, the translation function is important both within and across the various phases of the project. In some instances translation is accomplished by a tool or technique that, once implemented, becomes an automatic part of ongoing design and development processes. Other translation practices, more dependent on the informal efforts of individual team participants, may need constant managerial support if that capability is to have ongoing strategic significance. Since many of the "communication problems" in product design and development can be traced to failures in translation, the leverage that can be achieved from building a powerful portfolio of technologies and practices in this area is enormous.

The competitive strategy of an organization should be a guide as to the most critical aspects of translation to develop and the level of resources that can be justified toward this end. Translation enhancements can lead to improvements in both the effectiveness and efficiency of NPI projects. Improved effectiveness results from translation initiatives that stimulate the creation of better products. Efficiency is enhanced by routinizing communication and thereby reducing errors from incorrect translations and oversights from incomplete translations, which could require costly and time-consuming subsequent revisions.

Focused Information Assembly

At various points in the NPI process, certain design tools and techniques may bring multiple sources of information to bear on a particular problem, thereby leading to greater choice and improved decisions. This may be accomplished by routinely providing certain information earlier in the project than otherwise would occur. Another type of improvement along the same line is to assemble sets of related information that are traditionally available at the same time but not in a consolidated form facilitating easy access and use. We call this information processing function, in any of its manifestations, "focused information assembly."

The practice of involving vendors early in the NPI process naturally leads to earlier knowledge of vendor capacities, capabilities, costs, and constraints. Having manufacturing involved early in the product design process encourages the use of existing parts, processes, and tools whenever appropriate and the establishment of effective purchasing relationships. As discussed in the appendix, both quality function deployment (QFD) and design for assembly (DFA) techniques provide a framework for collecting and integrating multiple sources of information. Similarly, design reviews by groups composed of diverse technical and other specializations provide either formal or informal occasions for focused information assembly; such practices serve to aggregate varied and innovative perspectives on many aspects of design, tooling, and service. Manufacturing systems simulation requires collection and integration of detailed product and process information; this technique can provide valuable feedback for the improvement of both product and process designs. Activities of these types may also support the project by providing new awareness of existing information and more extensive usage of it.

Focused information assembly is especially valuable in support of nonroutine design activity where participants with different functional backgrounds must work together. Such activity is often enhanced when novel interpretations and decisions arise due to the mix of people involved. Here, synergies in interpretation arise from collaborative consideration of issues (more good ideas). For example, a designer and manufacturing engineer, either co-located or joined electronically by CAD and

CIM technologies, can simultaneously apply two broad but often different sets of knowledge to a problem. In this case, the manufacturing engineer brings knowledge of manufacturing equipment capabilities and product material trade-offs relative to manufacture, while the design engineer brings knowledge of desired product functions and general material abilities to achieve those functions. They may simultaneously optimize design and manufacturing, but often more importantly, bring whole new ideas and approaches to resolution of the problem at hand. Focused information assembly thus leads to greater choice and improved decisions.

Through such initiatives, collaborative decision making may encourage the anticipation of potential problems, early resolution of problems or even to resolution of problems that would otherwise not be resolved at all. Admittedly, as discussed in Chapter 4, it may be difficult to establish a workplace culture where effective collaboration in product design and development is widespread. Still, a necessary step in this direction is the focused assembly of information.

Communication Acceleration

The timing of access to information is particularly critical to a project's success because it can directly affect the cycle time to introduce a new product. Design technologies and practices that allow downstream stages of the process to start earlier (such as simultaneous engineering) may also reduce chances of oversight and error (and thereby reduce variability in the process), and allow evaluation and selection of greater numbers of alternative approaches to the particular task. Such technologies and practices achieve the function of communication acceleration.

An example of communication acceleration in the context of simultaneous engineering is the usage of computer-aided design (CAD) coupled with group technology (GT), electronic data interchange (EDI), and cellular manufacturing. This set of technologies and practices helps vendors start early design of parts and tooling, reserve production capacity, and order materials earlier. Such computer-integrated manufacturing (CIM) technologies can greatly enhance the communication across functions. Similarly, the provision of a tentative bill of materials for the new product assists manufacturing planners in formulating human assembler technical skill requirements and in turn supports hiring and training, all accomplished earlier than normal.

The development of early product prototypes, even in the concept development phase, is another example of communication acceleration. Here when design engineers provide an initial set of product specifications, the development of a preliminary prototype is made possible even before specifications are refined or formally approved. Also, technologies allowing early, rapid prototype development ensure that a physical unit or part is available early to those who will work on other downstream activities (such as tooling or packaging). Conducting product and process development activities earlier than normal can yield insights valuable to the ongoing design effort.

Communication acceleration practices can also facilitate the portions of new product introductions that are carried out by field service and marketing/sales groups. Field service needs to train field engineers in new methods and perhaps develop new diagnostic equipment. Marketing/sales needs to develop promotional material and educate key actors in the product distribution network. Providing appropriate product design information to these groups as soon as it is available will allow them to get an early start in activities that are critical to a successful product release.

A common benefit of communication acceleration technologies and practices is the potential for early development of needed skills and experience. Consider, for example, what can be done when manufacturing engineers are given early access to information on the production of a new product by participating in the assembly of engineering prototypes. This early production experience can lead to the development of better manufacturing tools, thereby reducing the extent of manufacturing start-up problems. Similarly, early knowledge of part specifications allows vendors to start production early and thus proceed down the learning curve. In both of these examples, communication acceleration ends up supporting quicker ramp-up to full-volume production.

It should be clear from these examples that communication acceleration contains an element of risk not present in the information processing functions of translation or focused information assembly. Here, in the interest of reduced product development cycles, a company may intentionally release information earlier than prudence would dictate. The goal is speed to market, which can have tremendous competitive significance. The cost, however, may be wasted effort from such "early starts." In general, the pursuit of reduced product development cycle times is likely to be sufficiently important to make communication acceleration an important information processing function. However, managers need to appreciate that the potential risks—in terms of increased product cost and reduced quality and market acceptance—argue against the single-minded pursuit of speed, as discussed in Chapter 3.

Furthermore, managers must not confuse a strategy of competing through the continuous improvement of products, with a strategy of increasingly accelerated product introductions. The former relates to the rate at which successive projects are completed (or started), while the latter deals only with the speed of executing a single project. Either strategy can be pursued independent of the other, or they might both be pursued at the same time with a combination of initiatives.

Productivity Enhancement

The efficiency with which some required design or development task is conducted can be improved directly by adopting certain design technologies and practices. Such enhancement of NPI productivity can be significant for its own sake, independent of improvements in the other kinds of information processing functions already described. We define these productivity enhancements to be the use of technologies or practices that improve the speed or reliability with which one or more of the routine activities of new product development takes place. Much as productivity has long been viewed as an objective in the field of manufacturing, this same perspective can be applied to the process of product design and development. Admittedly, the narrow-minded pursuit of productivity improvement in this arena can raise strategic dilemmas similar to those in the field of manufacturing, such as the loss of flexibility through excessive refinement of tasks, roles, and equipment (Skinner, 1986). Nevertheless, it would be a mistake not to acknowledge and selectively pursue the productivity payoffs that are available.

A drafting-oriented CAD package, for example, allows for quicker (relative to traditional manual methods) retrieval, drawing, and redrawing of parts and schematics. The CAD package would also provide other related and required documentation for manufacturing and purchasing purposes; for example, tasks that if not computer supported would have to be done by hand. This tool not only speeds up accomplishment of certain required tasks, but also reduces errors and supports more consistent and reliable information. Likewise, computer-aided software engineering (CASE) techniques can improve the productivity of the software designer, who is increasingly a key participant in the development of a wide variety of products and systems. At the broader level of project planning and scheduling, critical path methods and techniques (e.g., PERT) and other computerized project management tools may facilitate enhanced productivity by the product development team, and also reduce project planning and related management efforts. Computer-aided engineering (CAE) allows quicker rudimentary testing and analysis of proposed designs. Group technology in its many forms supports retrieval of old product and process information, reducing the need to "recreate the wheel." Computer-aided manufacturing (CAM) equipment, which automates robotic and numerical-control programming activity, can save time and minimize coding errors.

Analytical Enhancement

Certain analytical tasks are conducted as a matter of course in any traditional new product development project. Some design technologies and practices enlarge the feasible analytical range by providing unique capabilities to support the work of design engineers, product strategists, or other participants in the project. Such analytical enhancements are akin to those achieved by manufacturing tools such as job shop simulations or learning curve analyses that provide higher level, sophisticated information not available before. Similarly, these are applied to the design of new product or process capabilities. Tools for analytical enhancement and simulation make it possible to assure in advance that the new product will be more consistent than otherwise with some of the requirements of customers.

A CAD software package for conducting finite element analysis, for example, allows designers to consider thermal gradients and other physical properties to a degree that might not otherwise be possible at the early stage of designing a new product or part. Results from such analysis increase the effectiveness of the new product development, and can lead to reduced product cost, greater product reliability, and higher functionality.

Design engineers are also charged with developing "robust" product designs, which enable the product to perform its intended function well, even in extremely unfavorable environmental contexts and operational modes. By using statistical design experiments, simulation, and optimization models, and other statistical methods, design engineers can identify design parameters that cause product performance to change very little despite a wide range of potential environmental and use conditions.

These statistical techniques in aggregate are often called *Taguchi methods*. The practical payoff from this analytical enhancement is that the other more influential, or sensitive, parameters are assigned appropriate design and manufacturing tolerances, thereby improving the resulting product reliability. For the customer to receive the benefits from such robust designs, very high-quality conformance is required in manufacturing.

Management Support

Certain design technologies and practices facilitate the monitoring, assessment, and evaluation of a project, either in its entirety or with respect to specific issues. We call this the management support function because it serves to ensure a more thorough and systematic NPI process. Some tools, such as Gantt charts, PERT charts, and other kinds of project plans, help ensure that minor and major steps in the process are not neglected, guaranteeing greater process completeness, and adherence to sequence. Other techniques, such as a "contract book" specifying at an early phase of the project which organizations will deliver on specific aspects of the new product and associated production process, facilitate the use of common targets and a shared vision for the project.

As already described in Chapter 2, other practices, such as the formal performance reviews conducted in the phase-gate approach to product development, reduce the risk that is inherent in any new product introduction effort. Such aids to management support may promote both the efficiency and the effectiveness of new product introductions.

Where product cost is a serious basis for competition, one has an incentive to establish procedures and methods for confirming that materials and manufacturing costs are consistent with early assumptions. This includes attention to special equipment, tools, and fixtures. Where product performance is especially important, demonstrations at various stages of the development process (e.g., engineering prototypes and pilot runs), coupled with state-of-the-art test capabilities, are common. Assessing the designs for test equipment is particularly critical in those industries where the test equipment can be more complex than the product which needs to be tested. Formal procedures for the qualification of key suppliers is also important in a growing number of industries where conformance, dependability, and reliability requirements are reaching new heights. Design-for-manufacture checklists, requiring design and manufacturing engineers to carefully consider each part for manufacturability, and manufacturing sign-offs on product designs to guarantee manufacturing's willingness to accept responsibility for producing the product are also in widespread use.

Practices of this sort are a necessary but costly and time-consuming part of controlling the progress and outcomes of an NPI project. The challenge is to establish a cost-effective set of managerial supports. While good news (no problems) is always desirable, the real purpose of such approaches is to generate bad news (anticipation of design or development problems) as early as possible.

Managerial Perspectives

The six information processing functions described previously can be thought of as the life support systems of a company's new product development capability. Regardless of the complexity of the product, the size of the company, and the number and skill level of its people, these information processing functions are all essential. The six functions can be grouped into two sets, each promoting a different type of capability: (1) cross-functional integration; and (2) an efficient and effective NPI process.

The value and difficulty of cross-functional integration, within the product design and development process was discussed in earlier chapters. Such integration is accomplished through translation, focused information assembly, and communication acceleration. The common theme across these three information processing functions is furthering communication in ways that link related streams of design and development activity. These three functions thus promote meaningful collaboration among the members of a product design and development team, a challenge that was explored in Chapter 4. Furthermore, the same three information processing functions that promote cross-functional integration among team members also provide operational capabilities for implementing simultaneous engineering (described in Chapter 2).

The translation function accomplishes cross-functional integration by converting critical information from the viewpoint of one specialist to that of another. Focused information assembly brings together, in both time and place, information that supports improved design and development decisions. Communication acceleration operates in the time dimension, bringing vital information to participants sooner than they would otherwise have it. This set of functions reduces the level of "noise" in complex networks of design and development decisions. It promotes a common vision for the project effort as a whole and it strengthens the connections across the contributions of individual participants and subteams.

There may be some trade-offs among these three cross-functional integration functions. For example, a strong emphasis on focused information assembly will occasionally delay the timing of certain key decisions which, in turn, may offset other gains in project timing due to initiatives in communication acceleration. There are likely to be some synergies as well. For example, focused information assembly coupled with communication acceleration may facilitate more rapid skill development and smoother transfer from design engineering to manufacturing: many companies are creating initial production organizations, combining both design and manufacturing engineers, with this goal in mind.

Cross-functional integration has been identified over and over again as an important ingredient of effective product design and development. Now we see how certain design technologies and practices may promote such integration, both within the NPI team and between team members and others who need to become involved on a limited basis.

Productivity enhancement, analytical enhancement, and management support initiatives (in contrast with the other set of three information processing functions) improve the efficiency and effectiveness of an NPI project by affecting how it is conducted. The first two of these functions directly accomplish this through the introduction of new technologies and practices: productivity enhancement through the reduction of costs (often by using automation tools); and analytical enhancement through the achievement of desired product capabilities. Management support practices also promote efficiency and effectiveness, but in a less direct manner, by establishing information bases, procedures, and structures for more constructive review and control over the project. Design technologies and practices associated with these three information processing functions can be justified in terms of specific contributions that they make to the efficiency and effectiveness of the project.

The three kinds of integration functions affect NPI efficiency and effectiveness in less transparent ways, as they promote smoother and more tightly linked interpersonal activities. Admittedly, integration functions can be seen in retrospect to improve aspects of efficiency and effectiveness, but they are not so clearly associated with such outcomes in advance.

A critical trade-off in these functions arises between productivity enhancement and management support; if attempts at management support get overly bureaucratic, the associated loss of time can easily offset any gains received through productivity enhancement technologies and practices. The opportunities for synergies across these functions is also significant; technologies and practices for management support can promote the adoption of technologies and practices for both productivity and analytic enhancement.

Efficiency and effectiveness in the process of product design and development will also affect the achievement of the targets set in Phase 1 of the project. Again, we see how design technologies and practices may provide needed capability to the team and, ultimately, enhanced value to the customer.

SUMMARY

The design and development of new products is information intensive. It requires accumulation of data and insight from diverse sources and different functional perspectives. As a new product is being designed and developed, information is constantly accessed, interpreted, augmented, transformed, and deployed. The accuracy, consistency, and availability of such information essentially determines the extent to which a new product achieves required functional capabilities, ease of manufacture, and a fit with overall strategies. Adopting an information processing perspective on new product development facilitates the strategic assessment of different design technologies.

As product design and development becomes an increasingly complex process, management needs to pay closer attention to the potential value of alternative supporting technologies. The scope and variety of tools and techniques available to help people create and validate designs for new products and associated production processes is already staggering. Some of these are "soft" technologies, commonly called *methods* or *practices*. Others are software packages, sometimes relying on elaborate data bases, or networking capabilities. For all of these design technologies, and others yet to be introduced, management must ask: What benefit will they produce? Is it worth the cost and effort to implement?

Selecting and implementing design technologies and practices for use in the introduction of new products is more of a corporate bet than a guaranteed investment. Driven by the external forces of market opportunity, technological progress, and competitive pressure, manufacturing companies review available design technologies and practices, hoping to choose those that will help produce a potent strain of competitive advantage. Unfortunately, such choices often lead to confusion, unreasonable expectations, or misdirected effort.

Six different information processing functions are central to the successful design and development of new products: translation, focused information assembly, communication acceleration, productivity enhancement, analytical enhancement, and management support. Different design technologies and practices can enhance each of these functions. By paying more explicit attention to these functions and ways of achieving them, companies can promote cross-functional integration and create an NPI process that is more efficient and effective.

These information processing functions affect not only the outcome of single NPI effort, but also the ability to sustain a successful pattern of NPI outcomes. It is this pattern which either fits or does not fit a company's competitive strategy. The descriptive model presented in this chapter promotes an understanding of the information processing functions of NPI and their relationship to different design technologies and practices. As with many of the conceptual foundations presented in the preceding chapters, no single set of design technologies is best for all companies, nor is there a fixed, optimal sequence for adopting them.

APPENDIX
DESIGN FOR ASSEMBLY (DFA) AND
QUALITY FUNCTION DEPLOYMENT (QFD)

This appendix describes the techniques and practices of design for assembly (DFA) and quality function deployment (QFD), both of which are particularly rich in their possible impacts on product design development. Each of these is being actively considered for widespread adoption in different industries. Their methodologies, modes of application, and their expected outcomes are presented with links to the information processing functions (see Chapter 6) which they tend to strengthen the most.

Design for Assembly

Design for assembly (DFA) is a specific version of the general class of approaches to help ensure that products, as designed, are easily manufacturable. DFA, in other words, is a subset of DFM (design for manufacturability). DFA, in particular, is a systematic analysis process primarily intended to reduce the assembly costs of a product by simplifying the product design. It does so by first reducing the number of parts in the product design, and then by ensuring that the remaining parts can easily be assembled (Boothroyd and Dewhurst, 1987). This close analysis of the design is typically conducted by a team of design and manufacturing engineers, although other functional expertise such as field service and purchasing may also be included. DFA is used for discrete manufacturing products, and primarily for durable goods, but occasionally for consumer products. Since it is used to optimize assemblies, it is often used for smaller and medium-sized products, or for many subelements of larger systems. DFA does not specifically support system level applications and is usually applied to subassemblies (Stoll, 1988).

Whether conducted manually or with software, DFA techniques lead to a simpler product structure and assembly system. DFA algorithms build on many earlier industrial concepts including group technology, producibility engineering, product rationalization, and time and motion studies. In many ways DFA is a structured, automated approach to time and motion industrial engineering, combined with a bit of design philosophy via design axioms and guidelines (Andreasen, Kahler, and Lund, 1983).

There are two usages of DFA. It may be used to redesign a product already in manufacture (or a product being "remarketed," or reverse engineered). In this case, the product is disassembled and reassembled with special consideration of parts handling (feeding and orienting) and attachment (insertion) times and costs. These times and costs are found in data tables, through software, or by empirical observations. DFA may also be used for analysis of a product while it is still in the design phase.

DFA was developed with the assumption that the bulk of manufacturing costs are set in the design stage itself, before any manufacturing systems analysis and tooling development is undertaken (Nevins and Whitney, 1989). The primary objective of DFA is to minimize parts counts, thereby having fewer parts to be manufactured and assembled, fewer parts that can fail, and fewer interfaces between parts. It is these part interfaces that contribute primarily to product failure by providing sources for failure (Welter, 1989). The second objective is to have remaining parts of a nature that they are easily assembled together (Boothroyd and Dewhurst, 1987).

DFA provides a quantitative method for evaluating the cost and manufacturability of the design during the design stage itself. Some firms may choose to apply the analysis later in the design stage, and others may do so quite early such as during initial concept evaluations. In either case, the analysis steps require that an initial design is developed or proposed first. Then this design alternative is assessed penalty points for each feature of the design. These points when aggregated help determine the "design score" efficiency of assembly for the design, based on the expected material and assembly costs. Then the product is "redesigned" (this can be conceptual only) using part and product level design rules coupled with consideration of annual volumes and existing manufacturing processes. This step requires engineering creativity because DFA, even with rules, guidelines, and measures, is still an art. A typical design guideline is achieved by software queries asking these questions for the case of two parts connected by a fastener: "Does the fastener part move? Does it have to be a different material from the two parts? and, Does it have to be removed for servicing?" If the DFA team's response to all three questions is *no*, then the software would advise the team to make the assembly as a single part, thus eliminating two parts (Andreasen et al., 1983).

DFA has also served as a tool for supplier selection and involvement. This aspect is highlighted as large firms continue to move design and assembly functions to suppliers. Ford requires certain vendors to evaluate their designs with DFA before they can submit bids (Kirkland, 1988). DFA can also provide leverage over suppliers since firms can estimate product costs better.

Depending on how it is used, DFA can serve one or more of the following information processing functions described in the previous section. To begin, DFA provides *translation* capabilities from product design configuration to assembly cost, exact parts, and resultant equipment and personnel needs. It also provides important *analytic enhancement* because the DFA algorithms enable new kinds of subassembly design analysis that precede the development of physical prototypes and generate early findings that might otherwise not arise.

DFA serves other information processing functions in a more secondary manner. Its relationship to *information focus* is that the practice of DFA requires the bringing together of particular product configurations, general assembly costs and times, design axioms, and manufacturing systems principles, all to support a design choice. When used in early design or prototype analysis, DFA provides advance information to the manufacturing organization regarding needed equipment, labor skills, and parts. In this fashion DFA supports the firm's *communication acceleration* function. Another contribution of DFA is in the realm of *management support*: This practice enforces desired reductions in parts count; provides a basis for resolving issues of manufacturability; strengthens supplier management; focuses engineering efforts, and provides a relative evaluation score for alternative designs.

The collective strategic impacts of DFA can be considerable. DFA is both a tool and an organizational mechanism, or—in the terms of Chapter 6—both a design technology and practice. Since DFA requires teams of some sort, previous experience with cross-functional interaction, data collection, and analysis supports DFA usage. Alternatively, DFA helps build such experience. DFA may be seen either as being equivalent to simultaneous (or concurrent) engineering, or as a subset or a precursor to it. The organization that uses DFA improves and learns as many individuals work together concurrently collecting information, documenting development efforts, and applying improved bases for decision making. In the product literature of Boothroyd Dewhurst Incorporated, a Xerox executive is quoted as saying that when engineers use DFA software "they become more proficient in understanding the product delivery process so that each design becomes better and better." Ford uses DFA as part of their simultaneous engineering efforts, stating "it goes a long way to break down the barriers that exist between groups" (Welter, 1989).

DFA requires that design and manufacturing engineers coordinate activities, but does not specifically require the involvement of others. Its actual value in serving the information focus function may thus be less than is ideally achievable. For example, while IBM's Proprinter II is lauded for a fourfold increase in reliability, it is not considered to be easily serviceable; in this case, field service had limited involvement in the DFA team (Bebb, 1990). DFA can be used to integrate supplier involvement in the NPI process, but has limited ability to involve customers (though some may be on DFA teams for usability input).

DFA increases the time and resources spent in the design stage of the NPI, but cuts time throughout the rest of the development process. It does so by ensuring only one iteration of design and manufacturing, by reducing tooling acquisition and development requirements, by reducing training needs, by involving manufacturing and supplier functions earlier, and through the many timesaving benefits derived from simultaneity. DFA does not necessarily increase the very up-front product planning and system level configuration planning, but does take more time when tangible subassembly design alternatives are in hand.

Quality Function Deployment (QFD)

QFD is a systematic approach for the design of new products or services based on close awareness of customer desires coupled with integration of corporate functional groups. This approach is a set of planning and communication techniques that focus and coordinate organizational capabilities in order to develop products that most closely meet customers' actual needs. QFD can be seen in two lights: (1) as a comprehensive organizational mechanism for planning and control of new product development, or (2) as a localized technique to translate the requirements of one functional group into the supporting requirements of a downstream functional group. Most U.S. usage is of the localized variety, and so we focus primarily on this type of application (Akao and Kogure, 1983; King, 1989).

QFD requires a multifunctional team of experts to participate simultaneously in a complex analytic exercise, bringing diverse perspectives to bear on key issues of design. In localized usages, QFD serves both technical and organizational functions. Primary expected outcomes of QFD usage are: (1) increased *awareness of the customers' desires*, improved product with focus on actual customer wants, improved product specification setting, and, in general, more successful and effective products; (2) synergistic gains resulting from *integration* of and increased effective communication between individuals from different functional groups, such gains include understanding of each other's functional constraints and capabilities and transfer of information earlier than otherwise normal; and (3) effective *definition and prioritization* of new product development activities leading to increased understanding of trade-offs, clear understanding of product objectives and customer needs, improved decision making, a better end product and reduced product development time (due to elimination of remedial design iterations).

The most common example of QFD is translation from customer desires regarding improvement of an existing product into actual design engineering changes to be brought about in the next version of the product. This application is often referred to as the "House of Quality" due to the shape of the data matrix used to accomplish the translation (Hauser and Clausing, 1988). Customer demands are determined in a customer's own terminology via various market research techniques, and are ranked in terms of relative importance. Such a demand might be "ease of writing flow" for a pen. Design characteristics (such as "pen ink viscosity" or "pressure on ballpoint") are then "correlated" with each customer's desire to determine the degree to which it achieves the customer's requirement. Certain design characteristics may complement or offset each other; these trade-offs and opportunities are noted. Customer perceptions of competitors' products are generated to scale the advantages and disadvantages of the firm's product. Priorities and target values are set in an attempt to satisfy customer desires through particular design characteristics (e.g., have low ink viscosity and high ballpoint pressure). An example of the "First House of Quality," linking customer requirements to product design requirements, is shown in Figure 6–2. The "Second House of Quality" would have product design characteristics translated into required manufacturing processes to support production of the product (Smith, 1990).

QFD revises the time and intellectual energy commitments at each stage of the NPI process. In traditional NPI projects, the concept development phase may be somewhat short, with long periods of time devoted to such activities as design, manufacturing implementation, and redesign iteration. QFD greatly increases the time and resource requirements during the early planning/conceptualization phase, but if successfully implemented, reduces resource requirements for downstream stages.

FIGURE 6–2
QFD Illustration: Lead Pencil Example

A-1 Pencil	Length	Time between sharpening	Lead dust generated	Hexagonality		A Rate of importance	N Company now	Plan Competitor x	Plan Competitor y	P Plan	Ratio of improvement	B Sales point	C Absolute weight	D Demanded weight
Easy to hold	O/42			O/42		3	4	3	3	4	1	1	3	14
Does not smear		O/69	◎/207			4	5	4	5	5	1	1.2	4.8	23
Point lasts	△/44	◎/396	O/132			5	4	5	3	5	1.25	1.5	9.4	44
Does not roll	△/19			◎/171		3	3	3	3	4	1.33	1	4	19
Total	105	465	339	213	1122							Total	21.2	100
Percentage	9	41	30	19	99									
Company now	5"	3pgs	3g	70%										
Competitor x	5"	5pgs	4g	80%										
Competitor y	4"	2.5	3g	60%										
Plan	5.5"	6pgs	2g	80%										

(Left side labels: Customer demands (what) / Quality characteristics (how))

Main correlations:

◎ 9 = Strong correlation.

o 3 = Some correlation.

△ 1 = Possible correlation.

$$D = A \times B \times C \qquad B = \frac{P}{N}$$

Source: Reprinted from *Better Designs in Half the Time*, Figure 4.1, p. 4-3. Copyright © 1989, GOAL/QPC, 13 Branch Street, Methuen, MA 01844. Tel: 508-695-3900. Used with permission.

The most common U.S. approach to QFD uses the House of Quality method to trace customer requirements through to manufacture. This approach is quite effective for improving the cost, reliability, and functionality of existing product subsystems and parts. It is of limited use, however, for the design of entire, complex systems or those with dynamic product concepts. This limitation in the localized QFD application is due to a dependence on existing company data bases to reflect prior system level product concepts and familiar technological options for the product and the manufacturing process. Furthermore, American

QFD does not generate entirely new product ideas and is therefore frequently coupled with implementation and usage of the Stuart Pugh (1981) new concept selection method. This method forces engineers to develop and evaluate many new concepts, and so promotes fresh thinking. It is used as an input to the QFD process (Clausing and Pugh, 1991).

QFD serves to remove nonvalue-added activities by avoiding redesigns, by streamlining communication and reducing errors. It focuses energies on prioritized activities, and so serves as a method for planning and control of design and other specific tasks, leading to more efficient and less variable processes. QFD can act as a mechanism to make sure options and steps are not forgotten, again reducing chances of downstream problems. Through QFD, creative energies are less likely to be squandered on activities that will not eventually benefit the product. Instead, QFD adds value to the product and NPI process by better coordinating information and people to better meet customer requirements and increase organizational capabilities.

QFD's primary role is as *translator of information*, in the form of requirements, first from marketing to product engineering and then to manufacturing. QFD also clearly serves as an *analytical enhancer*, as it facilitates systematic data collection, comparison, and decision making. Through the application of QFD, many experts come to understand both the new product and the new product development process to a much more complete degree. The customer voice is injected throughout QFD activities, and so engineers and others see design characteristics in a new light.

QFD serves to focus information assembly by encouraging unhindered and informal cross-functional communication, data sharing, and problem resolution that would not otherwise happen. A major side benefit of such team building is the avoidance of bureaucratic structures and barriers to communication. Because many of the participants in a QFD exercise would normally not participate in the NPI process until a later stage, this technique also acts, secondarily, as a *communication accelerator*. In addition, QFD serves a *management support* function, especially when implemented on a large scale, since alternatives and trade-offs must be explicitly considered, and decisions must be agreed upon by a group.

Sustained usage of QFD can lead to a number of beneficial strategic outcomes. Among them are: faster new product time-to-market, early and/or faster starts for subsequent product introductions, availability of continuous supplies of new products to customers, and quicker compilation and updating of complete product lines. QFD usage in certain applications can lead to improved product quality, reliability, and serviceability. A favorable reputation for customer response will arise, as will significant product differentiation. On the organizational side, QFD leads to development and usage of a highly effective communications base within the company, and supports organizational learning through enhanced documentation, sustained interfunctional relationships and improved coordination. QFD is also a foundation for establishing an NPI rhythm, the special environment in which product development is a routine activity with a planned frequency. However, QFD may not support the frequent adoption of the newest product and process technologies because this technique is most rigorous when building on data from past product design and development projects.

PART 2

CASE ILLUSTRATIONS OF THE CONCEPTS

The chapters in Part 1 described the complexity inherent in the process of product design and development. The effectiveness of this process depends on a wide range of contextual factors regarding a company's market, competition, technology, structure, culture, work force, and strategy. Accordingly, it is difficult to appreciate the richness of this subject solely from a conceptual point of view.

Part 2 uses in-depth actual company illustrations of product introduction projects to promote an integrated understanding of the subject in different strategic and operating contexts. Each of these cases was selected to illustrate aspects of the topics covered in Part 1 (and to motivate the consideration of other topics covered in Part 3). Each case deals with a different company, industry, and market. Each also has other dominating contextual factors and yields insights about their significance. The cases serve to avoid the dilemma of "the blind men and the elephant," in which communication about the whole is hindered due to a preoccupation with individual parts.

Taken as a set, these case illustrations emphasize that, with respect to product design and development, there is more than one kind of elephant. This part of the book, therefore, is designed to help readers find similarities and differences between organizations they know and those that are presented here. This type of selective "benchmarking" is essential if people want to learn about their own company's issues, based on the experience of others.

These four case histories differ somewhat in presentation style as well as substance, as each was originally prepared by a member of the company that is the subject of that case. While this might be somewhat distracting at first, the reader is urged to consider matters of style as additional "data." The relative emphasis on topics in each case—and even the language used—reflects the business unit, the particular product design, and the development project that was studied.

Just as one gets a distinctive feeling of the enterprise from walking through different factories, so does one sense variety in the form and spirit of product introduction efforts across companies, or business units. Significant differences are highlighted in the commentaries following each case history, along with a review of general lessons suggested by the cases. The cases themselves, however, are purely descriptive and are not intended to be models of either good or bad practice of product design and development. Read them, instead, as illustrative, actual syntheses of conceptual material presented in Part 1 and as preparation for Part 3—which looks beyond the bounds of a single project at enablers, linkages, and the continued pursuit or more effective product design and development.

CHAPTER 7

PRODUCT DESIGN AND DEVELOPMENT AT A START-UP COMPANY: NeXT CORP.*

In September 1985, Steven P. Jobs used his own financial resources to found NeXT, Inc., with five former associates from Apple Computer. His vision was to design an affordable computer for advanced applications in higher education. Jobs and his team had an extraordinary technical and managerial background for such an effort, but this group would have to contend with an intensely competitive computer industry with rapidly changing markets, technologies, and product offerings. The small company had teamwork and group decision making as its hallmark from day one, and continued this philosophy and practice into collaborative product design and manufacturing development, and into relations with their few carefully selected vendors.

This chapter describes critical events and decisions in the design and development of NeXT's debut product, with special emphasis on the period of 1986 to 1989. A review of the early history of this start-up company provides examples of many of the topics presented in Chapters 4 and 5. Understanding the approaches taken by this resource-rich and sophisticated start-up is instructive for its own sake. Thinking about new product design and development in a small company is also a useful basis for comparison with the three large-company cases presented in Chapters 8, 9, and 10.

PRODUCT ORIGINS AND COMPANY CULTURE

In the first year the founders of NeXT devoted much of their effort to review of available and potential technologies in light of academic computing needs, and in the fall of 1986, formally launched their new product

---

*This chapter (excluding the author's added commentary section) is an edited version of the original case study prepared by Liza Gentile and Stephen R. Rosenthal (1989).

development effort. Based on their perception of the state of the art at that time, they set as their initial target the development, within two years, of a "three-M" machine (1 million pixels of screen resolution, 1 million bytes of main memory, and the ability to process 1 million instructions per second) that would retail for $3,000. The product concept would change greatly over the next three years as a result of evaluation of changing market needs and the desire to push technologies to their limits. Computing workstations decreased in price and personal computers gained in performance, both adding to NeXT's competition. Also, the three-M specification objective was soon expected to be quickly outdated in terms of industry product performance. Even with the dynamic nature of the design intent, most of the fundamental attributes such as appearance, data storage technology, chip design strategy, and microprocessors were defined in the first 12 months, and subsequently held firm. Industrial design consultants aided in conceiving a unique, functional, and ergonomic product package.

NeXT's culture strongly shaped by the vision of its president, was a dynamic one. Built by a handful of talented individuals who embraced the opportunity for tremendous reward and contribution at the risk of public failure, NeXT offered every member of its strategically lean staff (approximately 275 employees as of June 1989) the opportunity to learn, grow, and impact the organization. Jobs explained that teamwork and integration are necessary in an environment where precious human resources are used to their fullest extent: "Business has gotten so complex and dangerous and quick-moving that no matter how brilliant people are at the top, it's not good enough any more . . . you have to utilize every brain in the whole organization."

Given this belief, "quality begins with quality people" was dogma at NeXT, and great care was taken to staff accordingly. A painstaking interview process supported the company's efforts to import bright, energetic, skilled individuals eager for assimilation into a group-oriented culture. It was anticipated that most of the first 500 employees would bring 5 to 10 years of industry experience to their positions at NeXT. These employees would continue to be recruited as individual contributors, but their prior technical and managerial experience would facilitate their growth into management positions at NeXT.

While NeXT's growth was monitored by a meticulous hiring practice, employees described the internal workings of the organization as open and supple. Decisions were broached in a group format by staff members who were expected to study issues, formulate opinions, and subsequently present and defend them. (Consensus management did not prevail in the face of executive authority, however.)

Jobs' cabinet of functional managers, known simply as *Senior Staff*, met every Tuesday to troubleshoot and plan while monthly Company Staff assemblies provided a forum for the presentation of topics that were relevant to NeXT's extended population. Periodically, the entire organization convened off-site for an event that had come to be widely known as a *retreat*. Characterized by several days of meetings, presentations, and informal gatherings, these outings hoped to facilitate both vertical and horizontal communication throughout the corporate population.

Vice President of Manufacturing, Randy Heffner, defined the start-up's corporate mission as threefold: (1) the design and sale of great products, (2) the cultivation of an environment where outstanding people are encouraged to explore their potential while developing a product, and (3) the dispersion of wealth among those responsible for its generation.

CONCEPT DEVELOPMENT

The metamorphosis of NeXT's first product from an economical personal computer in 1985 to a personal workstation in 1988 was a function of market research coupled with gained knowledge of technological potential: As viewed in retrospect by Jobs:

> We wanted to build an incredibly neat computer for education, and we were going to work with education to do it. And what we thought it was when we started was the neatest computer we could think of, and then we had our eyes opened to a much neater one. I suppose you could call it a change of vision, but it didn't seem like that. It seemed like, first you're juggling with two balls, and all of a sudden someone throws you a third, and you say "oh, ok, I can manage this." Then they throw you a fourth, and you say "ok, I can manage this," and it's still juggling—only with more balls in the air. That's what it felt like.

Customer Involvement

The process of shaping this product began with the formation of an academic advisory board, and while this committee's insight was valued and sought as a vaccine against "group think syndrome," its members did not dictate the design of this product. Instead, their suggestions were solicited and molded into a foundation from which NeXT worked to create. These educators and researchers were cautiously apprised of decisions and changes once NeXT perceived them to be realities, and the motive for this selective disclosure was twofold. First, NeXT wished to avoid carelessly presenting considerations whose implications had yet to be fully explored, and second, employees feared sharing confidences with outsiders as external pressure to reveal the start-up's closely guarded secrets mounted.

NeXT executives recognized that the canvassing of market input would challenge their ability to isolate what customers really wanted from among those features they professed to need. While many members of the user community are technologically fluent (particularly within the world of academia), many are not, and are subsequently unable to project their perceived requirements into the distant future.

Timing Considerations

The task of weighing risks and alternatives in anticipation of keeping the machine as interesting and contemporary as possible during the development cycle was equally engaging. Short time-to-market was seen as a powerful competitive weapon in an industry where windows of opportunity routinely snap shut during a design's gestation. NeXT also understood that efforts to achieve short product development cycle times are frequently thwarted by an overwhelming temptation to chase impending technology forever.

NeXT's strategy for coping with this potential treadmill of perpetual design evolution involved preserving the stability of those product characteristics (such as basic software architecture and VLSI—very large-scale integration—chip design) that required long development periods and could not be recycled in the face of critical shipping dates, while constantly refining or enhancing other aspects of the design as circumstances allowed. NeXT executives were aware that an ill-considered feature selection could prove disastrous for a start-up with perhaps one opportunity to win an increasingly demanding and fickle market's favor. Accordingly, the design choices that supported NeXT's goal of creating a revolutionary computer that would have lasting industry impact were often made without any clear sense of what was "right," but with a gut feeling that a certain approach was worth "risking the company."

Formative Decisions

While NeXT's computer grew in scope, the workstation market exhibited rapid growth and turbulence during the company's first three years of operation. A crossbreeding of the once distinctive personal computer and workstation concepts to beget a hybrid generation of "personal workstations" offered competition from both segments of the market. User-friendly PCs were growing more powerful while expensive workstations were growing more affordable, and as the gap between their prices narrowed, they both began to compete for the same desktop computer market.

Most of the product's fundamental attributes were defined in the first 12 months by the founders, including the choice of external shape and appearance as conceived by Frogdesign president Hartmut Esslinger, the incorporation of the magneto optical disk-drive technology as pioneered by Canon, the preference for Fujitsu-routed VLSI components over multiple discrete devices, and the implementation of Motorola's three processors. During the embryonic stages of NeXT's growth, George Crow and Richard Page functioned as materials managers, and the corresponding evolution of a strategic pledge to maintain both a limited supplier base and lean part count was influenced by the small number of key people who had formative vendor responsibility. Page explained, "on the CPU board, I was literally handling all the interfaces with all the vendors. And there's only one me, and a bunch of them, so the simple solution is to keep the numbers down."

DESIGN/MANUFACTURING INTERFACE

Codevelopment of the product and process was fundamental to both the R&D and manufacturing strategies. NeXT was committed to the cooperation and integration of these engineering divisions. This was relatively easy to accomplish for this small entrepreneurial company, with no existing conventional attitudes or practices to combat.

The addition of manufacturing representation to NeXT's embryonic staff in early 1987 was strategically anticipated to benefit the development process. Norm O'Shea, Process Engineering Manager, and Kim Spitznagel, Automation Engineering Manager (both numbering among NeXT's first 50 employees), were hired with the expectation that they would contribute to this endeavor immediately. O'Shea recalled that "within a week we were expected to know what was on the board, what the components were, and who the suppliers were—and have some strong opinions about it." At issue was the choice between two different approaches for attaching components to printed circuit boards: (1) through-hole insertion and (2) the use of newer, more advanced surface mount technology. Jobs described his motive for establishing this group during the relative infancy of the formal design effort as a belief that attention to manufacturability is critical, and must commence at the outset of product development:

> When you're making printed circuit boards, the layout of the parts is really important, so if you lay out the board for through hole—and at the last minute go surface mount and re-lay it out entirely—it's kind of like taking 20 million steps backwards because during the process of laying it out for through hole you may revise the board ten times, and each time you improve it a little. You get none of that accumulated experience—or very little of it—when translating to surface mount. What you want to do is start with the technology you're going to ship with, and do all the revisions in that technology so . . . all that experience just rolls right into the product. . . . The worst thing in the world would have been to switch over to surface mount at the last minute. It's much better—much less painful—to go with surface mount right at the beginning of the process.

NeXT started with a vision but had no existing manufacturing capability (i.e., facility, equipment, and workers) and no network of suppliers. From the beginning the NeXT computer and manufacturing process were developed simultaneously. Integration between test engineering and R&D was highlighted as a strategic necessity. A mandate for designing a computer with a small number of parts influenced all design decisions, including the use of VLSI semiconductor technology. Careful development of a limited number of qualified suppliers was also a primary goal. Other manufacturing constraints were stated clearly early in the process, and adhered to. They conducted much of the development using CAD/CAM (computer-aided design and computer-aided manufacturing, respectively), and the organization as a whole was supported by a fully integrated computer information system.

Constraints on Product Design

Management believed flexibility to be essential to an integrated design effort, and correspondingly avoided prioritizing functionality, performance, low cost, and manufacturability. They acknowledged that the establishment of a rigid order of precedence might have resulted in the satisfaction of one requirement to the exclusion of the rest. Constraints were introduced to the design process, however, that allowed for maximum standardization without impeding creativity.

Manufacturing constraints which might have limited the product designers included: uniformity of board width, tooling hole locations, fiducial locations (focus markers for the vision equipment), robot part-type capacity, and a preference for surface-mount over through-hole components given the relative reliability and speed of that equipment.

The mandate of "few parts" had tremendous impact on the NeXT computer system design. In order to achieve a high level of functionality without proliferating part count, NeXT aggressively pursued VLSI design. This policy of "integrating on silicon rather than fiberglass" eschewed reliance on multiple, discrete devices connected by paths on the board in favor of consolidating the electrical interconnects for a wide range of functionality in large, complex chips. Each of the two large VLSI chips on the board replaced 75 to 80 individual components. As a result, while similar computers required two or even three boards to achieve comparable performance, the NeXT computer was driven by only one.

Careful maintenance of a limited supplier base complemented this endeavor to minimize part count. While many established companies were actively pruning installed supplier bases, NeXT was cautiously shaping theirs with the help of rigorous approval mechanisms and high-quality standards.

These dual objectives of minimizing part count and supplier number frequently opposed one another in practice. Grundy explained that although his group was committed to these goals ("in digital design, we just about cry when we have to put another part on the board because we really believe that if there is any possible way not to put another part on, we should do that"), several factors could render the addition of a new component preferable to supplier base proliferation. For example, should an engineering change order (ECO) introduce the need for a 200 ohm resistor—and the board currently uses only 100s and 50s—it might be advantageous to incorporate two 100 ohm resistors into the design if board real estate allows rather than source a new component, or fill another station on the robotic equipment whose part-type capacity is limited by this factor.

Management of Product Design

When manufacturing was co-located with R&D during the early stages of NeXT's development, communication and joint problem solving in the tiny Palo Alto facility was readily managed: a simple trip up or down one flight of stairs provided answers to queries about product and process criteria. Despite a company commitment to providing every employee with a computer and access to the network as a means of bridging the geographical rifts between NeXT's dispersed properties, this group's relocation 25 miles away to Fremont (California)—and subsequent reliance on electronic mail, phone calls, and visits to communicate with their former neighbors—forced consideration of a more formalized interface to ensure that critical exchanges between these teams continued unhindered.

An ECO policy dictating that designs remain unaltered unless proposed changes were documented and approved with a signature by all designated personnel typified NeXT's efforts to implement standard protocols governing the design/manufacturing relationship. This formal process was further enhanced and supported, however, by unofficial practices that guided team efforts to apprise counterparts in other functions of their intentions at the earliest opportunity.

Digital engineering would, for example, call purchasing the day they even suspected that they might want to add a new part to a board in order to alert that organization to this possibility. Verbal communication was often accompanied by a FAX to the factory, and the goal of this anticipatory contact was the allowance of maximum lead time to source a part, and correspondingly, the opportunity for purchasing to indicate that supplier issues might adversely affect Digital's proposal. If the idea was pursued beyond this preliminary exchange but was not readily accepted as proposed, an ECO meeting was organized and attended by representatives from R&D, manufacturing, purchasing, and marketing/sales. Not only did this meeting serve as a forum for the discussion of suggested changes, but the opportunity to advocate desired enhancements was exploited as well. (Manufacturing issues which might arise to this end included part placement for improved robot yield, solderability, and packaging concerns.) At the meeting's close, a list of authorized changes was generated, minutes were taken, and the modifications were planned.

Engineering Prototypes

Following redesign, an engineering prototype run of perhaps 10 boards was scheduled to confirm the viability of these enhancements. Barring any disappointments, a formal ECO for production of the new board was drafted electronically in a "was/is" format, and immediately dispatched by E-mail to affiliated staff. Hard copy was eventually routed to these same individuals for signatures, but the content was expected to be familiar to its recipients given this preliminary review practice.

"Continuity and communication" were identified as the cornerstones of success in this endeavor. NeXT employees recognized that concurrent information flow between Palo Alto and Fremont was paramount to the practice of integrated design, and the company's endorsement of liberal idea exchanges suggested this process. Continuity referred to engineering commitment to production designs. In order to fully exploit a product's potential following its release to manufacturing, there exists a considerable afterlife of design enhancements that require prolonged engineering attention.

Both Haven and Miskell attended a weekly "problem parts meeting" in Fremont, and agreed to extend this time to include a sustained engineering meeting pertaining to the design and assembly of the cube. If a manufacturing problem was identified in the agenda, the group would actually adjourn to the factory floor to evaluate the situation. (Typical issues that warranted review include supplier modification of purchased material, fixture and tooling changes, and actual design changes.) Essentially, the objective of these meetings was product/process improvement as the result of combined final assembly and mechanical design effort. Plans may have existed, for example, to redesign a final assembly tool, but discourse with mechanical engineering might have revealed that a simple design change would facilitate the use of existing tools at far lesser cost.

NeXT set up their manufacturing facility in Fremont, a 40-minute drive from corporate offices. NeXT had maintained co-location of design, manufacturing, and test engineers up to this point, and while actual co-location ended, close communication was still maintained.

Test Engineering

In the face of swift component functionality and complexity progression, integration between test engineering and R&D was highlighted as a strategic necessity. The challenge of problem isolation is amplified when a complex chip (such as the VLSI devices which are capable of a wide range of functions) fails in a multitasking environment. While writing special algorithms that enable the software to check the hardware assists this endeavor, this practice represents a huge drain on resources, and was rejected by NeXT in favor of "designing the system from the bottom up to circumvent this" by actually building testability into the chips themselves.

For this reason, test engineering worked closely with R&D to establish testability standards for custom chip design; an engineer was hired whose sole charter was the definition of test requirements in chip design for the entire organization. While based in Fremont, this individual worked with his counterparts in digital design to promote testability.

PARTNERSHIP WITH SUPPLIERS

NeXT's supplier selection strategy and philosophy also supported an integrated design effort. In accordance with the company's desire to avoid a practice of "inspecting quality in," suppliers were chosen who demonstrated both the ability to ensure the quality of the materials they delivered, and the willingness to work with NeXT in a partnership to continuously improve the quality of the products their joint efforts achieved. The nature of this relationship was expected to be an open, dynamic one.

NeXT reserved the freedom to survey a supplier's production process or product design, and subsequently to offer suggestions for improvement. (Similar input from an educated supplier base was welcomed in reciprocation.) The goal of this exchange was the collaboration and pooling of talent, and was typified by examples of interplay between NeXT's design engineers and three of their key component suppliers.

Canon: Advanced Storage Device

Perhaps the single most significant risk the first 20 to 25 employees assumed in product design was the selection of the untried magneto optical disk drive as a primary storage device. The founders were introduced to this technology during a tour of Canon's labs in Japan when NeXT was in its infancy. After wrestling with questions of the drive's viability and appeal, they ultimately decided that the benefits of mass storage outweighed the risk of its implementation. While several other companies were actively considering this technology as backup for a Winchester, NeXT saw the potential for the 256-megabyte CD-like optical disk to replace floppies as the medium of the 90s and was the first in the industry to ship an MO drive in volume as a main storage device.

Despite enthusiasm for the concept, however, NeXT disliked Canon's cumbersome interface between the drive and the computer (two cables, 50 to 60 wires, and a large interface board), and subsequently worked aggressively with this supplier to simplify the design. The economical results of this modification effort—a single, 20 pin ribbon cable that facilitated assembly while enhancing reliability, and a chip that communicated exclusively with the drive—were designed in Palo Alto simultaneously with Canon's continued development of the drive in Japan. Grundy owned primary responsibility for this project, and reflected on the success of NeXT's relationship with Canon:

> I had my piece of paper on this side of the Pacific, and they had their piece of paper on that side of the Pacific, and in September a year and a half ago they brought their first prototype drive for me to check out. I had just gotten my new chip back three weeks earlier . . . and they flew out with their new drive, and the great thing was, within two days we had connected the board to the drive, and had it reading and writing. It's just a tremendous feeling if you're an engineer, to know that you followed all the rules correctly.

Fujitsu: Custom Chips

NeXT selected Fujitsu to manufacture their custom VLSI devices from a pool of three highly capable vendors based on the availability, reliability, and cost of this supplier's commercial package. NeXT assumed a substantial design risk in trading supplier manufacturability for product performance, and subsequently relied heavily on Fujitsu's proven track record and undisputed reputation as the industry leader in the manufacture of large, complex dies.

Several considerations that impact manufacturing must be addressed in VLSI design. The first is package selection. As circuits are condensed, heat dissipation increases, and specially designed packages that facilitate cooling are required to avoid overheating. There also exits a technological limit to the number of "gates" (electrical interfaces) that commercial packages can readily accommodate, and NeXT's designs challenged these boundaries.

A state-of-the-art package at the time these chips were facing production could accommodate a maximum of 20,000 gates, and yet NeXT was determined to force 24,000 gates of functionality into that limit. Fujitsu guaranteed its customers that, if silicon use of a standard package was less than 90 percent of the usable space (18,000 gates in a 20,000 gate package), they would route the design and manufacture it without difficulty or delay. Rather than sacrifice critical functionality and a commitment to minimized part count, NeXT opted to bend Fujitsu's rules, and submit a design for routing that required approximately 95 percent of the available silicon. This decision forced an undesired level of design complexity upon Fujitsu once the 90 percent constraint was exceeded, and generated pronounced manufacturing difficulties.

In effect, NeXT chose an alternative that slowed their time-to-market considerably while Fujitsu struggled to route their designs. Ultimately, however, the incorporation of these gate arrays supported both NeXT's vision of desired product performance, and in-house manufacturing strategy. The alternative would have entailed the substitution of many simpler components for the large, custom chips, thereby proliferating the number of parts that NeXT's robots would have had to place. Very large-scale integration was a key decision that had a major impact on manufacturing. By not going to an easier discrete design, which would have helped NeXT to get to market sooner, the company avoided taking on a much more complex manufacturing burden.

Sony: The Monitor Assembly

The striking monitor design was the fruit of group effort. It was initially crafted conceptually by industrial design consultant, Hartmut Esslinger. Subsequently, George Crow and the analog engineering group at NeXT translated this vision into a manufacturable reality that satisfied the specific needs of Sony's manual assembly process. Contrary to the American predilection for bulky, snap-together designs, Sony preferred working with parts that were joined with screws as experience with this method allowed them to achieve high levels of efficiency. Furthermore, storage requirements with this design were minimized. Large quantities of the flat metal sheets it required could be stacked neatly; Sony's product was space-intensive only upon completion when it was immediately packed and shipped.

This compliance on NeXT's behalf was motivated by a desire to accommodate their supplier's requirements, despite the potential for this conformity to haunt the designers. Should NeXT ever be forced to consider an alternative OEM (original equipment manufacturer) source for the monitor, the alliance with an organization that favors the rejected snap approach will force a renewed design effort to accommodate this change in process.

MANUFACTURING STRATEGY

The founders did not initially anticipate NeXT's evolution into a large-scale corporation, and correspondingly eschewed in-house production of the printed circuit board. At first, they planned only to do in-house assembly of the cube itself. Then they decided that in order to control inventory properly it would be necessary to build the entire CPU themselves. They then selected a site for their dedicated production facility.

Use of Surface-Mount Technology.   Manufacturing technology options were addressed early. While conventional through-hole insertion initially won favor, a thorough review of this technology coupled with an investigation of surface mount revealed the latter to be better suited to NeXT's vision of manufacturing in the 1990s as this process had the ability to place a greater number of parts with fewer, more reliable machines than required by the alternative. Subsequent to this decision, Kim Spitznagel and Norm O'Shea were hired at the beginning of 1987 to design the hybrid board line that integrated surface-mount and through-hole equipment as the result of large capacitor and connector packaging scarcity in the preferred format at the time of its design.

The founders' case for automation (and resulting commitment to surface mount) transcended the obvious quality benefits of building printed circuit boards "hands off." Printed circuit boards are sensitive to both mechanical and static damage, and while touching a board may not destroy it immediately, manual contact has the potential to prematurely degrade performance. Other more subtle implications of this strategy weighed heavily in their evaluation of manufacturing alternatives, however, and ultimately supported their decision in 1986 to embrace a technology whose implementation was identified as a significant risk for the start-up.

Automating for High Volume and Flexibility. Highly automated advanced manufacturing was anticipated to support NeXT's goal of building a quality product with reduced time-to-market, but it was further chartered with facilitating the establishment of a team capable of nurturing and implementing a forward-looking vision of the future. In the spring of 1989, NeXT's 40-member manufacturing organization was staffed with over 400 man-years of experience; more than 70 percent of the managers and engineers possessed advanced degrees. Jobs identified the correlation between the type of manufacturing concern a company creates, and the people it is able to attract to sustain it:

> If you want your manufacturing organization to be state of the art—if you want your people to be great . . . you have to allow them to be on the edge of technology. . . . They want to be where the action is, and where the action is right now is in surface mount and CIM (computer-integrated manufacturing)—and all those other things. If you don't do those things, you're not going to get the best people, and you're going to build an old way of manufacturing organization. . . . Part of what happens is people think that doing things like surface mount is risky. What they don't realize is the risk of not doing them . . . we've got a manufacturing team that is so good and so excited; I think it's the best I've ever seen.

Volume flexibility was identified as the key to success. NeXT realized that as a company grows and sales escalate, manual insertion necessitates rapid hiring and training. The ramifications of this sudden human resource proliferation include the potential for high turnover and a resultant erosion of quality control as the organization adapts to these fluctuations and employee inexperience. Automating dissolves these issues as the ability to ramp-up—or down—is not impeded by human resource realities. While manufacturing was staffed by 40 people, NeXT projected the ability to do $1 billion in sales with the addition of only 60 employees to this group.

On March 30, 1989, just 120 days after going into production, NeXT announced Businessland's commitment to purchase a minimum of $100 million worth of computers at wholesale prices for sale in their national chain of retail outlets. By then, approximately 1,000 units were installed at professional developer sights and universities, and production had been paced to enable careful monitoring of the board line at this level of output. Jobs had confidence in the factory's ability to accommodate the surge in production the Businessland alliance suggested, and cited NeXT's investment in flexible, reliable automation as the foundation for his faith:

> This Businessland deal is going to get us into higher volumes than we expected much sooner than we expected. If we hadn't put the infrastructure in place with the automation, there is no way we'd be able to do this. We couldn't have signed that deal in good faith, because we wouldn't have been able to make the product—maybe at all—and most certainly with the quality we desire.

All of the senior participants at NeXT knew that the ability to implement readily any needed product design changes would be integral to their success in a rapidly changing workstation industry. NeXT pursued this kind of flexibility by deciding to invest in general-purpose robots that were not product-specific, and by developing the internal capability for sophisticated software development as part of automating their manufacturing system.

With the exception of those robots designed in-house, all of the machines that were ultimately selected for the board line could have been replaced by three or four equally desirable pieces of equipment from other vendors. NeXT purchased early model state-of-the-art equipment that was "taken from the shelf," dismantled, modified, and reassembled by the process engineering team. O'Shea explained that this practice of integrating standard equipment into a process through redesign is a common Japanese strategy that is less prevalent in the United States, "American manufacturers tend to go back to their suppliers and beat them up if things aren't exactly what they expected."

The risk of this strategy, of course, was equipment failure. The NeXT board line was single serial; there was no provision for redundancy. If any one of the machines faltered, the entire line ground to a halt. O'Shea continued, "but I think it was a risk we felt comfortable with because we had enough experience knowing that, 'these are the things the machine must do, and we feel that we can overcome the newness of it.'"

SUBSEQUENT OUTCOMES

The NeXT computer system was introduced in October 1988, before manufacturing capacity was completely in place. It would be almost a year later before the first release of its operating system software. The computer system was initially offered through direct sales to the academic community for $6,5000. In March 1989, Businessland began retailing the workstation for $9,995. The NeXT computer had grown in scope from its original vision to meet the perceived changing needs of the workstation market.

NeXT failed to achieve commercial success with its initial product offering, apparently due to a combination of its price and shortcomings in its features and capabilities (relatively low-processing speed, black-and-white monitor, and general lack of software applications). It was reported that the original model was withdrawn from the market in May 1990, having sold fewer than 10,000 units. On September 18, 1990, Steven P. Jobs, NeXT's Chief Executive Officer, introduced four new computers, the company's second-generation products. In at least two of these models, the black cube was replaced by a smaller unit the size of a pizza box. The optical disk drive would now be an option and all new machines would contain floppy disk drives. The base list price for the new NeXT-station machines would be $5,000 (including a black-and-white screen and a 105-megabyte hard-disk drive) and there would be an $8,000 color version. The more advanced models were called NeXTcube (starting at $8,000) and NeXTdimension (selling for $14,000).

COMMENTARY ON THE CASE

For a start-up company, NeXT Corp. was unusually sophisticated and financially well-endowed and had a leader with an impressive and relevant track record. The founders had lived through the heady days of successful product design and development at Apple and they added other principals with extensive experience in other high-technology companies. Steve Jobs' personal fortune supported the venture and it was not long before they received massive infusions of external funds, most notably from Canon. Steve Jobs was already legendary as a high-tech entrepreneur and his new venture had to be taken seriously from the start. These factors all contributed to the viability of a new company that had yet to develop a single product.

Being small, NeXT had no need for a formal process of executive review for their initial product offering. All of the executives were involved in this effort on a day-to-day basis and there were no functional barriers to have to overcome. NeXT seemed well-positioned to launch an effective product design and development effort.

Planning and Managing the Project

From the beginning, however, NeXT faced formidable obstacles. They started with a vague image rather than a specific product idea. They were jumping into the rapid waters of what turned out to be a highly dynamic and competitive new market: the personal workstation. Their visionary leader wanted the first product to be radically different from other products likely to be available within the same time frame. They set out to become a world-class manufacturer in order to be able to offer product quality that would exceed even the best of the competition. Without an existing product to benchmark, NeXT intentionally took a leap into the future, where both market and technology risks can be considerable.

Developing Collaborative High-Performance Teams

The youthful enthusiasm, boundless energy, mutual admiration, and shared reward structure that characterized this company from the outset provided a workplace culture highly conducive to the joint problem solving that was required. This workplace culture, as suggested by the mood as well as the substance of this case history, is notably different from that of most larger (and many smaller) companies.

NeXT seemed to do well in sustaining needed collaboration of various kinds during its early years. Industrial design resources, attained from a highly regarded external consultant, were used effectively at an early stage to create an attractive image for the company and its debut product. The design of product testing was closely tied to the design of the product itself. From the beginning, the company invested heavily in the creation of a CIM (computer-integrated manufacturing) capability. This early commitment to building a very advanced and flexible manufacturing capability made process design and development as high a priority as the design of the product itself. Communication between design and manufacturing remained close, even after the manufacturing group moved to their new plant location. An ECO (engineering change order) policy provided needed discipline even though the workplace culture was highly informal.

Coping with Technology Choice and Risk

NeXT had a vision for their debut product and they designed a manufacturing process that suited this product. After rejecting initial decisions to outsource their printed circuit boards, NeXT purchased proven standard manufacturing equipment which their process engineers then dismantled and modified to suit their own particular needs. In-house production increased their control of this critical element.

The company took major but calculated technological risks in product design and were very selective in their use of suppliers. Partnerships with world-class (Japanese) suppliers moderated the many inherent risks in developing the advanced workstation product that they had in mind. In selecting the untried magneto optical disk drive as a primary storage device, NeXT engineers worked aggressively with Canon to simplify the design of an interface between the drive and the computer. In an intentional time/quality trade-off, they pushed Fujitsu to accommodate more than formerly acceptable VLSI density, thus allowing there to be only a single printed circuit board. This decision caused delays but simplified NeXT's manufacturing process and made the product more reliable. The mandate for few parts was intentionally ignored when the supplier of the computer monitor, Sony, adopted a design that had points joined with screws rather than a snap-together approach.

In their initial product introduction effort, NeXT took deliberate risks based on their strategy for success in a very competitive market. Surface mount technology was chosen for its forward-looking properties; engineering advances on through-hole equipment had fundamentally slowed, and surface-mount robots were often several engineering generations ahead of their more conventional counterparts. Even more important, however, surface-mount technology was identified as the most effective, efficient means of expeditiously producing high-quality products at the lowest possible cost. Although this decision contained a certain measure of risk, the resulting automation of the board line provided considerable flexibility for the future.

Note that NeXT (unlike the other case histories in the following chapters) built a new factory as part of its NPI effort. This posed opportunities in the sense that the entire manufacturing process (and culture) could be designed to satisfy the production needs of the new product. However, the associated challenges of doing all this within the time constraints of the NPI project were considerable.

Conclusion

In retrospect, it is easy to fault NeXT for failing to reconcile their product specification with a realistic business plan. Their initial product was late to market, due to the complexity of the technology (software and hardware) that it embodied. And when it was commercially available, the product was deemed to be too expensive. There were simply not enough customers interested in paying for some of the advanced capabilities of this computer workstation. At the same time, the initial product lacked certain relatively basic features that customers had grown to expect. If there is a lesson here, it probably is in the realm of questioning initial product specifications in terms of commercial assumptions about the market.

It is hard to fault this company for designing a product with unproven technology because this technological innovation is the norm, and product life cycles are shorter than product development cycles. The only way for a company like NeXT to successfully enter such an industry is to count on some of tomorrow's technology, for much of today's technology will no longer offer a competitive advantage.

NeXT appears to have learned from its initial commercial mistakes. It now has a product line that appears well-attuned to the marketplace. It continues to benefit from its superb manufacturing capability for high-quality, advanced printed circuit board production and outstanding partnerships with suppliers in product design and development (as well as ongoing production). Its collaborative culture continues to be a major corporate asset.

Although it is still too early to project the likelihood of success of NeXT, the early years of this company provide considerable insight into important aspects of product design and development. Whether it has missed its opportunity to be a commercial success in the highly competitive workstation market, only time will tell. External factors will no doubt be important in determining the outcome. But so will this small company's capability for organizational learning.

CHAPTER 8

A LARGE COMPANY TAKES A
NEW APPROACH:
NORTHERN TELECOM*

In 1976, Northern Telecom Ltd. (NTL), was the first company to introduce a fully digital central office switch, which revolutionized the telecommunications industry. Since then, NTL has worked aggressively to maintain a leadership role in the introduction of innovative telecommunications products to support the growing demand for advanced networking solutions. As of 1990, NTL was the world's leading supplier of fully digital telecommunications equipment (e.g., telephones, small-business systems, private branch exchanges, transmission equipment, and central office switches) in more than 70 countries. It was the second largest supplier of all telecommunications products in North America and fifth worldwide. Its major competitor within North America is American Telephone & Telegraph (AT&T), while its overseas competitors include Siemens (Germany), Ericsson (Sweden), and ITT (United States). In 1990, the corporation employed approximately 50,000 people worldwide and its revenues exceeded $6 billion.

This chapter chronicles Northern Telecom's development of the Meridian Norstar business telephone system during the period of 1985 to 1988. A review of this project provides an integrated sense of the topics contained in Chapters 2, 3, 4, and 5. Viewed in its entirety, the Norstar case history illustrates challenges faced by a large company in attempting a dramatic change in their process of product design and development.

ORIGINS OF THE PRODUCT

In 1984, Northern Telecom identified a substantial business opportunity in the market for 2-to-100 station small-business telephone systems. The pricing flexibility provided by deregulation of the U.S. telephone indus-

---

*This chapter (excluding the author's added commentary section) is an edited version of the original case study prepared by Jerry Dehner (1990).

try had prompted small-business consumers to shop for deals on telephone equipment. This demand had fueled a proliferation of low-cost products, often manufactured offshore, that were aimed at the small-business market. However, these consumers had quickly found that the cheapest products were, in the long run, not always the lowest cost products. Installation and maintenance expenses, as well as opportunities lost due to system downtime, often eroded the nominal savings gained through a low-purchase price. Small-business customers were becoming more value-conscious when buying voice-communications equipment.

Under the auspices of the General Manager of the Calgary Plant, who had manufacturing and marketing responsibility for the small-business market, Northern Telecom acted quickly upon identifying this business opportunity. It was clearly time to develop a new product for this market. The company's existing small-business telephone system (Vantage) was plagued by poor quality, "me-too" technology, high cost, elongated NPI schedules, and cultural walls between Marketing, Engineering, and Manufacturing. While Northern Telecom possessed a majority share of its Canadian market, Vantage sales in the $4 billion U.S. market were under 1 percent. Although the company already had a product, the SL-1 private branch exchange (PBX), for the high end of the business market (for users with more than 200 stations), it seemed unlikely that these digital PBXs could be modified for small-business needs.

Using information already gathered on competitive products, the Marketing Department in Calgary tested consumer reaction to various combinations of product features through a series of 12 focus groups. The focus groups also provided Marketing with an understanding of the end user's communication process. These focus groups represented single-site companies using personal computers and employing 30 employees or less. Each focus group consisted of eight employees, including the company's key decision maker.

This market research pointed to an overwhelming conclusion; users desired a key system that was easy to install, use, and maintain. Although extended capabilities and multiple features were desirable, if the system features were difficult to access or use, they added no value from the customer's perspective. Simplicity became the positioning theme for the proposed small-business system product, ultimately named "Norstar."

A competitive analysis further defined the market opportunity:

1. The target segment (2-to-100 station) represented 60 percent of

the business voice-communications market—a potential 3 million lines per year.
2. Small-business end users and the distributors were dissatisfied with the quality-versus-cost trade-offs offered by the existing products.

However, the competitive analysis also revealed major challenges facing Northern Telecom:

1. The market was not expected to grow; gaining share would mean direct competition with established products.
2. Introducing and delivering the product to the market would require significant investment in new product development as well as administrative, marketing, and sales support.

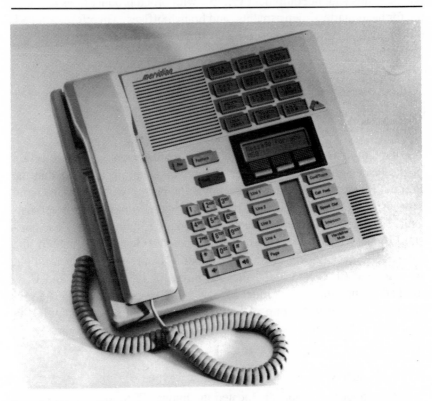

Courtesy of Northern Telecom, Inc.

3. Price competition limited the potential for substantial profits in the near term.

A computer-based simulation model was then developed as a "test product." This model allowed a panel of potential customers to have a hands-on experience similar to that of using the proposed new telephone system. By touching "buttons" on the computer screen as if they were making a telephone call, they could simulate different key strokes in a calling sequence. This simulation model facilitated the definition of preliminary functional requirements for the new product.

Northern Telecom's executive management decided for strategic reasons that the company should establish a major position at the small end of the business market. Without a complementary small-business product, it risked losing shares in the larger market, as some of its corporate accounts sought equipment vendors with broad product lines applicable for headquarters, departments, and branch offices. Therefore, after $1 million was spent on market research studies in 1984–85, approval was given to develop Norstar.

ORGANIZATIONAL APPROACH

A critical organizational decision by senior management was the selection of the leader or Product Manager of the Norstar project. (Note that the Product Manager at Northern Telecom is the person with full managerial responsibility for the success of the new product, while the Project Manager, as described later, had much more limited responsibility.) The Norstar Product Manager was the recently appointed General Manager of the Manufacturing Division in Calgary, Alberta, from where Norstar would be marketed (in Canada) and manufactured for the Canadian and U.S. markets. When the new Product Manager took over in late 1984, he said "I found a depressed group of inexperienced people in a money-losing business." He formed the following project team, reporting directly to him:

- Product Manager (located in Ottawa, Ontario).
- Engineering Manager (located in Ottawa, Ontario).
- Manufacturing Manager (located in Calgary, Alberta).

- Canadian Marketing Manager (located in Calgary, Alberta).
- U.S. Marketing Manager (located in Nashville, Tennessee).

The importance of the Product Manager was his ability to "champion" the project and foster a sense of urgency to change traditional new product introduction (NPI) approaches. Some of the statements that capture the personal philosophy of this Product Manager follow.

"Don't look at what we are doing today. Look at where we need to be tomorrow."

"Timing is more important than time."

"Good people will make a bad process work. Bad people will make a good process fail."

"Sometimes we make the mistake of trying to deliver technology instead of a product."

This Product Manager maintained top-management support during the entire Norstar development effort because he was a trusted member of their inner circle. This ensured the availability of needed financial resources to support the Norstar project. But this alone was not sufficient to ensure project success. The Product Manager led by example, thus fostering a more innovative approach to exploring alternative options and taking prudent risks. He also made critical decisions on the location and management of the engineering portion of the Norstar project.

It was typical for NTL's Engineering and Manufacturing organizations working on the development of a new product to be remotely located with respect to each other. For the Norstar project, Engineering was located in Ottawa and Manufacturing in Calgary, approximately 2,000 miles apart. This separation caused some delays in transferring information and making compatible and timely decisions, despite the extensive use of the company's electronic-mail network. Early on, this geographical dispersion was a barrier to developing personal bonding and mutual trust between members of the two organizations.

The decentralized nature of the organization often persisted to the lowest level in the organization. For example, it was not uncommon for the various engineering personnel (i.e., printed circuit board, integrated circuit, mechanical and software designers) to be remotely located, reporting to different managers and working on a variety of projects. In the past, this had caused considerable conflict over priorities and "ownership" into the projects.

Role of the Product Manager

The new Product Manager recognized the cultural and technical problems associated with this decentralized approach. While it was infeasible to consolidate Engineering and Manufacturing in one location, he was able to convince top management to co-locate all engineering functions in one location (Ottawa) and totally dedicate this engineering group to the Norstar effort. Moreover, this dedicated engineering staff reported to a single engineering manager whose only focus was the Norstar development effort. This consolidation under one manager helped to build quick ownership into the project by the engineering staff. This had great strategic importance to the ultimate success of the project.

However, negative short-term impacts on the project schedule resulted from this action. Consolidation in Ottawa caused an immediate deficiency in a number of engineering resources, particularly in the ability to obtain software designers. Since Ottawa is not noted as a "high-tech" area (such as California's Silicon Valley, Boston's Route 128, or Raleigh's Research Triangle Park in the United States), this location was a barrier to attracting needed technical skills. The unfavorable Canadian individual tax structure, harsher climate, and cultural differences, also made it more difficult to tap the larger pool of technical talent in the United States. Despite the shortage of needed engineering talent, a team of skilled professionals was assembled to design the system. The decision to consolidate and dedicate engineering resources to the project was a gutsy, strategic decision that facilitated the development of an innovative, high-quality product.

Another critical decision made by the Product Manager was the use of different engineering managers for the conceptual and detailed design phases of the Norstar development effort. The conceptual design effort was managed by an engineering manager with the technological skills needed to develop the innovative architecture for the system. Once the conceptual design was completed, the Product Manager brought in a new engineering manager whose strengths were focused more on project management and interpersonal skills to ensure good communication with other groups (especially Manufacturing) and rapid closure on the detailed (i.e., physical) design of the product. While the switch in engineering managers caused a project delay of approximately three months, this initial time lost was recovered by the excellent communication and closure achieved by the new manager.

Executive Gate Review

In 1984, Northern Telecom formalized an executive gate review process to assess the status and quality of all new product introduction projects. Formerly, the company was strongly engineering-driven and there was no formal series of reviews by senior management of a new product under development. The gate review process established formal linkages between successive stages of new product introduction and was designed to replace a technology-push approach with one that was market-driven.

This project was the first to employ fully the new executive gate review process, used to formally assess the progress at the end of each major phase of the NPI effort. This process was an extension of the design review process used in the Vantage product development and was instituted by executive management due to recurring NPI problems. As described in Chapter 2, the gate review process adds executive-level inputs and approvals at key transition points (e.g., design transfer to manufacturing) in the NPI effort.

Gate reviews were scheduled at the following transition points:

Gate 1: Between concept definition (marketing) and detailed product specification (engineering).

Gate 2: Between detailed product specification (engineering) and prototype manufacturing and test.

Gate 3: Between prototype manufacturing and test and production manufacturing and delivery to customer.

In the mid-1980s, Northern Telecom did not have a Gate 0, but the company subsequently recognized the importance of adding this initial gate to assess more formally the original product idea and its business justification.

The timing of the gate reviews is shown in Figure 8–1, along with other key project milestones. This project management process segmented the various phases of an NPI, with each phase ending with a formal, top-management review to ensure project objectives and deliverables were met. The project team had to successfully pass through a "gate" in order to continue to the next phase. At any of the gates, the project could be altered or terminated by executive management.

As one might expect, the transition to the new executive gate review process was painful. As an example, Norstar's passage through Gate 1 was first plagued by several delays (from December 1985 to March

FIGURE 8–1
Norstar Milestones

	January	February	March	April	May	June
1984			Start marketing study			
1985						1. Development team formed 2. Digital vs. analog decision
1986	Change engineering managers		1. Gate 1 failed 2. Test plan 3. Revised development plan 4. Canada marketing plan		Custom silicon pricing agreement	1. Gate 1 passed 2. Product specs completed 3. Final cost targets
1987				1. Original U.S. marketing plan 2. SMT decision		
1988	Production started		1. Gate 3 passed 2. Product launch			

July	August	September	October	November	December
		1. Marketing research study complete 2. Initial cost targets 3. LCD decision 4. Plastic KSU packaging	Project plan	1. Commercial specs complete 2. Development co-located in Ottawa	Initial development plan
	Canada marketing plan finalized		Manufacturing plan		
Tech trial		U.S. marketing plan implemented	Product name selected	U.S. marketing plan formally approved	1. Compact field test completed 2. Gate 2 passed

1986). This was due, in part, to the deliberate change in Engineering Managers in January 1986. Despite the additional time provided, the March Gate 1 review failed for several reasons:

Marketing had not yet defined its U.S. marketing plan.   Up to this time, Marketing had focused on the well-established Canadian market and had not begun to address the more strategically important U.S. marketplace at the time of the Gate 1 review. A separate U.S. Marketing group for Norstar was established belatedly in December 1986 to create and implement a detailed U.S. Marketing plan. This plan was developed in April 1987 and implemented in September 1987. Notably, as shown in Figure 8–1, this plan was not formally approved until November 1987, two months *after* its implementation. The U.S. Marketing group took the risk of starting "early" when the approval of the U.S. Marketing plan got bogged down and timing had become critical.

Engineering possessed insufficient software developers.   Engineering's main problem was trying to acquire the needed software developers to support the Norstar project. Since the engineering effort was consolidated in Ottawa, Ontario, it was difficult to ramp-up fast enough because Engineering could not realistically expect to tap the large U.S. software talent pool. Like many other advanced technology projects, the critical path in the Norstar project was centered around the software development effort.

Manufacturing did not complete its test plan.   Early in the project, Manufacturing worked with Engineering on the selection of appropriate technologies to be employed in the design of the Norstar product. This intense activity and the ongoing demands to maintain production on the Vantage product caused a delay in developing the Norstar Manufacturing and Test Plans.

The failure at Gate 1 was the low point of the project, since the "scrambling" to achieve closure on the gate review deliverables had resulted in the first (and last) major setback to the team. It took another three months (until June 1986) for these outstanding action items to be resolved and for Gate 1 to be passed. Gates 2 and 3 were passed without major problems in December 1987 and March 1988, respectively.

The gate review process ensured the actual closure of tasks and deliverables, rather than allowing the project to proceed while those that were "essentially" complete still required attention. Considerable activity by the project team was needed to gain executive-level approval to proceed to the next phase of the NPI process.

PROJECT PLANNING AND MANAGEMENT

The Norstar project proceeded from initial planning and to detailed design and development in accordance with the gate review process. This section describes Norstar's critical success factors and targets, the role of the Project Manager, the culture of the team, and relationships with suppliers.

Critical Success Factors and Targets

To be successful in the market, Norstar had to have these attributes: (*a*) ease-of-use and installation, (*b*) low cost, (*c*) availability, (*d*) high reliability, (*e*) upgradability, and (*f*) be feature rich (to include help screens, program keys, and special call handling capabilities). With such success factors in mind, product performance, time-to-market, product quality, and cost targets were set early in the project.

These targets helped Engineering broaden its responsibilities beyond functionality and performance of the product. The product performance, time-to-market, and cost targets were set as a result of the early, extensive marketing analysis. The product quality target was expressed in terms of the percent of volume production satisfying the specifications, or "manufacturing yield." A target yield of over 90 percent was set by Manufacturing at the time the other targets were established. "Mature first cost" (described in Chapter 3) was a key principle guiding all Norstar design and development decisions. Decisions regarding components, materials, and technologies were made to achieve target costs during the initial production run, without extensive design changes, learning curves, or production modifications.

Despite the need to achieve a record time-to-market target, Engineering was *not* squeezed for time. The clear lesson learned from the Vantage project was that reducing the time of the early NPI phases (e.g., Market Analysis and Engineering Design) simply increased the total NPI time-to-market by forcing more design iterations, since the product could not be designed "right" the first time. The Norstar project team understood the "pay me now or pay me (much more) later" options and elected to provide Engineering with the time that it needed to achieve a one-cycle design effort, with the caveat that manufacturability, testability, and installability, were part of the design deliverables.

Designing the product in one cycle eliminated the cost and time penalties associated with major design changes. Engineering quickly understood, through early and active manufacturing representation on the project team, the need for a design effort focusing on more than just functionality and performance. This early collaboration between Engineering and Manufacturing permitted concurrent design and manufacturing process development (i.e., simultaneous engineering) to become a reality, discussed later.

Role of the Project Manager

Norstar's Project Manager, a seasoned employee, had to ensure that the responsible line managers attended to the right issues, in accordance with the executive gate review process. With respect to product design and development, his role was more of an arbitrator and facilitator than a decision maker. He worked closely with Marketing, Engineering, and Manufacturing to ensure that decisions and deliverables were made by the appropriate team members on a timely basis. He also had ongoing responsibility to see that design and development activity was consistent with the time, cost, and quality targets that had been established early in the project. He conducted monthly project reviews, scheduled the gate reviews and initiated follow-ups to ensure closure of all outstanding action items. This Project Manager for Norstar, in summary, had many of the responsibilities that we have associated with a "Project Manager" throughout this book, although much of the business responsibility for the project rested with the Product Manager.

The Informal Organization and Team Culture

For the Norstar project, the informal (i.e., group decision) process was the key to NPI success in comparison with what transpired at executive gate reviews. The formal process was the vehicle to convey formal project status to executive level management and to force closure on key interdepartment deliverables. The informal decision process quickly evolved as a result of the very negative experiences on the Vantage project. Norstar was much more complex than Vantage from a technological, schedule, and cost-control standpoint. The formal and informal organizational approach helped all members of the Norstar team to "do it right the first time."

Over a period of time, fortunately, a strong daytime family culture developed between these traditional antagonists. Typically the various participants in an NPI at Northern Telecom had their functional goals to meet, setting the stage for cross-functional competition since they lacked a common basis for agreement. The early Norstar meetings were, therefore, naturally awkward for Marketing, Manufacturing, and Engineering due to the hostility generated by the Vantage project's failures and the traditional differences between Engineering's focus on applying the latest technology, Marketing's concentration on time-to-market and price, and Manufacturing's emphasis on producibility. A change in the culture began to occur only after these organizations began to understand the importance of the collective concerns of the team.

Partnerships with Suppliers

Strategic partnerships with internal and external suppliers was another key element of the success of the organizational approach. Internally, a unique pricing agreement with Northern Telecom's custom silicon manufacturing subsidiary in June 1986 permitted the lowest initial prices possible for proprietary integrated circuit chips, to be followed by higher long-term prices once the product had established itself in the marketplace. In this way, the internal Northern Telecom suppliers developed a vested interest in the success of the Norstar project.

External suppliers also became an integral part of the Norstar team. The cooperation between Northern Telecom and the suppliers of plastic telephone and system unit packaging began early in the conceptual design of the product. Northern Telecom lent its technical expertise to the plastics suppliers to help them improve their manufacturing processes and develop an electronic interface for design data transfer between Engineering and the suppliers. The impact of this collaborative effort was to ensure a reliable supply of high-quality plastic parts.

Northern Telecom also collaborated with the Universal Company, which supplied most of the automation machines used by Manufacturing to support the Norstar process. Manufacturing had a high-comfort level working with Universal to develop the Vantage flow line. In order for

Norstar to be successful from quality and cost standpoints, through-hole manufacturing yields from the automation equipment had to be improved from 500 parts per million (PPM) defects to 100 PPM. Manufacturing and Universal collaborated to recertify Universal's through-hole automation equipment to the 100 PPM nominal yield level. This process benefited both companies, since it permitted Manufacturing to achieve its needed through-hole yields while helping Universal expand the market for its improved machines.

ENGINEERING AND DESIGN

The design engineering process employed in the Norstar project was much different from the process used in the prior Vantage NPI effort. The Vantage engineering effort had been a classic case of "throwing the design over the wall" to Manufacturing, only to have it thrown back due to poor manufacturability, testability, functionality, and cost. This resulted in many frustrating iterations before the design was close to being acceptable to Manufacturing.

It was obvious that the objectives of the Norstar project could not be achieved by repeating the mistakes made during the Vantage NPI effort. The pain experienced by all organizations involved with the Vantage product development effort was the single most important catalyst for change. The frustrations and disappointments of the repetitive Vantage development effort had the greatest effect on the Norstar Marketing and Manufacturing groups, since these individuals had firsthand experience with the Vantage debacle. It is likely that the level of cooperation and teamwork demonstrated by the Norstar project team would not have occurred if it had not been for the frustrations of the prior Vantage NPI effort. These frustrations, coupled with the new Norstar product manager assigned to lead the project and the introduction of the new executive gate review process, increased the desire (and acceptability) for changing the NPI process and culture.

Comparisons with Prior Project

The numerous design iterations in developing Vantage had a very negative impact on the time-to-market, development cost, and quality of that earlier product. As an example, when first released to the market, Vantage circuit boards had over 120 barnacles. (A barnacle is a design change not formally integrated into the manufacturing process. It refers to a manual operation, such as a wire or an extra component, that is apart from the automated assembly process.) These barnacles were needed as a result of repetitive testing, which contributed to Vantage's increased time-to-market (over four years), reduced manufacturing yield (approximately 70 percent) and poor overall U.S. market share (under 1 percent).

The engineer's criteria for a successful Vantage design was limited to the traditional factors of functionality and performance. Because the Engineering organization did not assume responsibility for designing manufacturability, testability, cost-effectiveness, and robustness into the product, the Vantage NPI effort degenerated into a tennis match in which Engineering and Manufacturing kept returning the project to the other. It required another year after the Vantage product was released to resolve the more critical quality problems uncovered by customers in the field. This hurt the company's reputation in the small-business system market, particularly in the United States, and required a much stronger effort to ensure Norstar's success.

Beyond the increased time-to-market, development costs, and product costs, there were serious quality problems which plagued the Vantage product, after it was released in the marketplace. Despite these problems, the product enjoyed moderate success in Canada where it had an almost exclusive hold in the centralized Bell Canada marketplace. However, in the United States, this "me-too" product with its associated quality problems and high cost could not compete against foreign competition in an open, decentralized market.

Figure 8–2 compares the time-to-market and number of barnacles for Vantage and two versions of Norstar.

By comparison, the Norstar NPI effort demonstrated much better communication and cooperation between Engineering and Manufacturing, resulting in only one barnacle on the circuit boards. This single barnacle was removed within six months after the initial release of Norstar. More importantly, only one design change (repositioning the liquid crystal display) was needed during the entire Norstar NPI effort.

In summary, the Norstar NPI improved upon the Vantage experience in several dimensions: faster time-to-market (approximately three years), higher first-pass manufacturing yield (over 90 percent) and increased U.S. market share (9 percent in the first year). First-pass manufacturing yield is made up of circuit test (90 percent yield), functional test (98

FIGURE 8–2
Comparison of Norstar and Vantage Introductions

System	Time-to-Market*	Number of Barnacles†
Old Vantage	48 months‡	120+
Norstar Compact	21 months	0
Norstar Modular	14	1

*Gate 1 to Gate 3.
†Defined in the text.
‡Does not include 12 additional months to redesign product after Gate 3.

percent yield), and system integration test (97 percent yield). The gain in market share is particularly important, since the U.S. small-business system market is distributed among many competitors.

Decisions about Technology

Decisions regarding technology selection for the new product exhibited a greater business orientation, with emphasis placed on assessing the impact of technological decisions on the NPI targets. Throughout the entire engineering phase of the project, Marketing and Manufacturing worked to ensure manufacturability, testability, and marketability of the product. The appropriate level of technology, not always the latest technology, was selected to optimize the combined targets for success. In the mid-1980s, it was quite unusual for Northern Telecom to be driven by a business, rather than a technological perspective on a new product.

One decision about technology, in contrast, went the other way— toward the more advanced alternative. This was the basic question of whether the Norstar system should be based on an analog or digital technology. During the first half of 1985, Marketing and Manufacturing debated with Engineering, with the former preferring the analog approach and the latter the digital. Marketing surveys of potential small-business customers had identified no perceived added value from the more advanced digital approach. An analog system, based on an enhancement of the one used for the existing Vantage product would be cheaper to develop, as well as to manufacture. Meeting the tight development and cost targets were critical for the success of Norstar since the market, particularly in the United States, was price sensitive.

Engineering, meanwhile, favored a fully digital replacement to Vantage, thereby addressing the data transmission, as well as voice communication, needs of small businesses. Engineering was interested in applying a newly developed digital messaging and signaling technology, and they had prior experience in designing digital systems for other product lines. Marketing and Manufacturing, however, were much more comfortable with the analog technology that they had used before.

This digital/analog debate was finally resolved by the Product Manager on a strategic basis. The company had a visionary zeal for applications of digital technology and so, on faith, he ruled in favor of that technological option. He concluded that cost reductions would have to come from other design decisions. The decision to make Norstar fully digital ensured compatibility with the company's portfolio of digital products, while differentiating Norstar, with its state-of-the art voice and data capabilities, from its competition. These capabilities were to have value to the customers, even though the marketing survey on this issue, as constructed, had not helped consumers understand the functional significance of this technological choice.

A decision to use plastic, rather than the traditional and more expensive choice of metal, to encase the key service unit of Norstar was aimed at reducing the cost of the product. Metal had been used to help shield the processor from electromagnetic interference (EMI), such as picking up local radio stations on the telephone, but a plastic case would have no such shielding properties. This early (pre-Gate 1) and somewhat risky decision forced the circuit designers to select and lay out components on the printed circuit board so as to control EMI at the source. Engineering, accustomed to assessing their own success in terms of the extent of technological innovation, agreed to take on this challenge. Marketing and Manufacturing preferred the plastic casing option for its cost reduction implications and the fact that it would not pose any additional manufacturing risks.

One other notable pre-Gate 1 design decision was to include a liquid crystal display (LCD) on all Norstar telephone sets, rather than the industry standard of having it only on the operator's set. The market analysis had clearly indicated the need for a simplified approach to selecting telephone features and the LCD feature promoted ease-of-use through message prompting. It also supported the inclusion of a number of advanced features. The installation of the LCD substantially increased the cost of the telephone sets but set Norstar apart from all competitive products. Despite the fact that the LCD was the single, most expensive component of the telephone set, the team agreed that it should be included in the product design.

Use of Electronic Data Interchange

Electronic transfer of design, manufacturing, and test information, already common in Northern Telecom's new product introduction process, was also used extensively for Norstar. Especially significant was the full electronic transfer of mechanical design information to the plastic suppliers for their use in producing the molds for Norstar's casings for the key service units (KSUs) and telephones. This design-based technological capability reduced the time for the vendor to create the plastic molds from many weeks to several days.

MANUFACTURING

The biggest changes that took place during the Norstar development effort occurred in Manufacturing. Sales of the Vantage product had continued to decline at an alarming rate. The new General Manager (i.e., Norstar Product Manager) of the Calgary Division was forced to reduce the size of his organization in early 1985 to match the reduction in Vantage sales. Everyone understood that the survival of the Division depended on the success of Norstar.

Strategy

The extensive marketing analysis completed prior to NPI start-up was critically important, since it established unique system requirements, aggressive cost targets, and early market windows for product development. In order to satisfy these needs, the manufacturing strategy was to:

1. Develop the manufacturing process during (concurrent with) engineering design.
2. Demand high-manufacturing yields (build it right the first time).
3. Ensure that manufacturability is designed into the product during the engineering phase.

In the traditional fashion, building the Norstar manufacturing process would have begun (as with Vantage) after the Gate 2 review had been passed. It was clear from the project plan, however, that a concurrent engineering approach, in which Manufacturing began its work before the formal approval at Gate 2, would be necessary if Norstar was to meet its target delivery time, the first quarter of 1988. In other words, without such concurrent (or simultaneous) efforts by Engineering and Manufacturing, the Norstar introduction would have been delayed approximately six months, missing the market window, with associated loss in sales and potential market share. The milestone chart (Figure 8–1) clearly shows the impact of concurrent engineering by noting the short time (January–March 1988) needed by volume manufacturing once prototype field testing and Gate 2 were passed in December 1987.

As previously discussed, the requirement for improving manufacturing yields was calculated once the product price was set in early September 1985 by Marketing. The 70 percent manufacturing yields being achieved on Vantage—due to the extensive number of design modifications (additional wires and other barnacles on the board)—were much less than would be required for Norstar to be profitable. Manufacturing yield requirements on Norstar were calculated to be over 95 percent, much higher than other small-business system products and, more importantly, higher than most other Northern Telecom products. The manufacturing organization quickly understood the impact of this requirement. They would not only have to be proactive with Engineering to ensure manufacturability was designed into the product prior to Gate 2, but they also had to change their basic views on acceptable machine tool yields and the risks associated with selecting new manufacturing technologies.

Transition

Northern Telecom had embraced a strategy of using the Norstar introduction to expand its share of the small-business system market. Yet to the Manufacturing Division, as the Norstar NPI was getting under way, there was a natural preoccupation with survival. While the Engineering group was totally dedicated to the Norstar project from the beginning, the Manufacturing and Marketing organizations in Calgary had to support both interim enhancements to the existing Vantage product while ramping-up for the strategic Norstar replacement. Not surprisingly, a survivalist mind-set quickly developed in Calgary, forcing changes in the level of risk-taking by the members of the Norstar project team. This was most evident with Manufacturing's sustained resistance to Engineering's desire to employ surface-mount technology (SMT) in the product's design. Given the high-manufacturing yields required, Manufacturing was very hostile toward SMT until confidence was gained from the recommendations and assistance provided by the SMT experts across the corporation in April 1987.

The early collaboration between Engineering and Manufacturing helped to ensure a relatively smooth migration from the existing Vantage production line to the new Norstar product. Whenever feasible, large portions of the Vantage production process were ported to the new Norstar process. In the end, approximately 60 percent of the total Vantage process was used on the independent Norstar line, which included nearly 80 percent of Vantage's through-hole manufacturing processes. However, a totally new material handling system and surface-mount technology process was needed to support the Norstar manufacturing effort.

While Manufacturing understood by Gate 1 (June 1986) the basic level of process-related improvements needed to achieve the ambitious yield targets, it was not until October 1986 that the manufacturing plan was developed. Manufacturing's time available to address the manufacturing plan was severely impacted by their time-consuming manufacturability thrust with Engineering and their ongoing Vantage production support activities. In short, Manufacturing was too thinly staffed between Gate 1 (June 1986) and Gate 2 (December 1987) to support both Vantage and Norstar, given the proactive culture needed for Norstar to be a success. Only through extraordinary individual efforts were the competing Vantage and Norstar goals achieved.

The SMT equipment arrived at Calgary in late December 1987 and was fully operational by February 1988. The more conventional through-hole process for attaching printed circuit board components was fully operational by mid-March, due to heavy coordination of three external vendors and the in-house tool software development required.

Test equipment was acquired in February 1988 after it was realized in June 1987 that the antiquated test equipment used to test the Vantage product was not capable of supporting the testing requirements and quantities required by Norstar. From a test perspective, the telephone-set printed circuit boards were viewed as a subset of the KSU boards. The test group selected the same Hewlett-Packard automated test equipment for all circuit board tests. Advanced open-field methods were used for flexible, on-line acoustic testing, rather than the less flexible and more costly sound chamber method, which would also have imposed flow restrictions on the manufacturing line.

These strong supplier alliances were a necessary prerequisite to achieving a just-in-time (JIT) manufacturing process for Norstar. Parts moved from the receiving dock directly to the point of usage, skipping the traditional process steps of incoming inspection and intermediate storage in the stockroom.

Simultaneous Engineering

The process for selecting the SMT equipment, based on the recommendations of the corporation's SMT experts in April 1987, was another example of using collective wisdom to solve strategic issues. While the decision was made to minimize the use of surface-mount technology, the SMT custom silicon still represented nearly 70 percent of Norstar's system functionality. Manufacturing quickly gained confidence in the recommendations, since they came from SMT manufacturing experts who, unlike Engineering, shared the same work experiences and values with the Calgary Manufacturing organization.

The early collaboration between Engineering and Manufacturing permitted a high degree of simultaneous engineering activities. Figure 8–3 shows the percent overlap achieved with Design Engineering in each of the major manufacturing areas. Overlap, in this case, is measured as the percent of an activity that had already been completed by Manufacturing, when management gave formal approval to proceed. This definition is consistent with the idea (presented in Chapter 2) that simultaneous engineering involves some risk-taking because downstream activities are started before the related upstream activity is completed and approved at the appropriate gate review.

FIGURE 8–3
Extent of Simultaneous Engineering in the Norstar Project

Manufacturing Area	Percent of Overlap with Design Engineering*
Surface-mount process	80%
Through-hole process	25
Final assembly process	50
Functional testing	80
Overall overlap	50

*Defined in the text.

The level of simultaneous engineering achieved between Engineering and Manufacturing allowed an early start to the enormous effort leading to the complete installation of the Norstar manufacturing line. The highest priority for manufacturing was to obtain the automated surface-mount equipment since first prototype circuit boards could not be manufactured without this equipment. The design of the automated through-hole manufacturing process was validated by visiting manufacturing divisions in Research Triangle Park Raleigh (North Carolina); London (Ontario); and West Palm Beach (Florida). The manufacturing process used in West Palm Beach was 80 percent compatible with the desired Norstar process. Similar to the approach used to select the appropriate surface-mount equipment, the Manufacturing group used the collective experience within the corporation to minimize the risks associated with choosing the right through-hole manufacturing process.

SUBSEQUENT OUTCOMES

The National Communications Forum selected Norstar as the best new product introduced by the telecommunications industry in 1988. Manufacturing, which had stretched to reach a production level of 200,000 telephones in 1988, met a production goal of 1 million sets in 1989. In 1990, industry analysts ranked Norstar one of the top five market shareholders in the American small-business telephone systems market. By then, Norstar was being sold in over 40 countries at a volume several years ahead of their own forecasted growth.

In retrospect, it was easy to identify the planned design attributes underlying Norstar's market success:

1. Fully digital design providing high levels of performance and reliability.
2. A rich set of features, easy to use, and low in cost.
3. Rapid delivery, easy installation, and readily upgradable.

Shortly after Norstar's successful entry into the marketplace, many switch designers who had been working on Central Office products asked to join the small-business products (Norstar) team. Before then, Central Office products had been considered to be the more exciting engineering assignment.

COMMENTARY ON THE CASE

The Meridian Norstar telephone system was a product conceived in adversity by a company familiar with the pressures of global competition. Senior management, in this case the General Manager of the Calgary plant, acted quickly and firmly in support of the idea of a new product for the small-business telephone systems market, to replace the disappointing Vantage product. This decision was aligned with their competitive strategy of offering a broad product line to meet the needs of business customers of all sizes.

Structuring the Work

The Norstar project was the first opportunity for Northern Telecom to apply its full executive gate review process to structure the phases of product design and development. This gate review process was very similar to the version described in Chapter 2, aside from the fact that there was no formal Gate 0 review. Interestingly, shortly after the completion of the Norstar project, the company acknowledged the importance of a formal preproject review of the new product idea and added Gate 0 to all subsequent NPI projects. Though lacking this initial gate review, Norstar benefited from an early and thorough market and competitive analysis that soundly validated the new product idea. When the project failed to pass the initial Gate 1 review, all participants had to realize the expectations for thoroughness, timeliness, and discipline in the conduct of product design and development.

Planning and Managing Projects

Goals and priorities, set early in the planning phase of the Norstar project, proved to be an important source of its success. This project illustrates how such goals or targets, as described in Chapter 3, can help in resolving subsequent design choices for the product and its means of manufacture. From the beginning, ease-of-use, low cost, simplicity in installation, and upgradability were paramount concerns for the designers of Norstar. Concept development for Norstar was aided by a computer-based simulation model designed for test use by a panel of potential customers. Reliability and functionality were also important and led to the acceptance of limited use of new surface-mount technology. An early emphasis was placed on achieving mature first cost, in reaction to the disappointing manufacturing experience with the predecessor Vantage product. This shared goal facilitated the early involvement of manufacturing in the concept development and design phases. Time-to-market, though important, was deemed of less priority than the broad range of quality considerations. Although simultaneous engineering practices led to dramatic reductions in time-to-market, the philosophy of doing it right the first time dictated that more time be placed on the early phases of market analysis and engineering design.

Two managers were influential in different ways in shaping the design and development of Norstar. The Product Manager, who had business responsibility for Norstar, maintained top-management support throughout the project, made the formative decisions to base the product on digital technology, to co-locate the engineering functions in Ottawa, and to switch engineering managers between the conceptual and detailed design stages of the project. The Project Manager, in contrast, was more of a facilitator during the course of the project and steered the project toward the consensus needed at each gate review.

Developing Collaborative High-Performance Teams

The conducive workplace culture of the Norstar team, called a *daytime family*, was credited with achieving the multiple goals of this project. Achieving this positive spirit was not easy since morale and traditions of cooperation were low when the project began. Although the mood of crisis probably encouraged more cross-functional collaboration than was typical, the aspects of project planning, organization, and management mentioned previously were also clearly important in promoting collective decision making. Partnerships with key internal and external suppliers also promoted broader kinds of collaboration. The high levels of communication and cooperation between Engineering and Manufacturing were especially important in achieving a product design that was readily manufacturable at the necessary high yields. The process of cross-functional decision making, including appropriate negotiations, was effectively handled.

Coping with Technology Choice and Risk

As the Norstar project proceeded from concept development through engineering design, choices were made in support of digital technology, plastic telephone and system unit packaging, limited introduction of sur-face-mount processes, and improved through-hole automation equip-ment. Each of these decisions was tied to the basic product concept and the associated time, cost, and quality targets. Each decision was made in consideration of the associated risks. The introduction of surface-mount technology at this plant also provided a learning opportunity that would be important for the future. In the Norstar project, Northern Telecom, a noted "technology company" managed to make a business orientation— with special emphasis on value-added to the customer—central to its product design and development effort. Risks and trade-offs were care-fully considered and the appropriate resources were allocated to areas that had been defined, from the beginning, as somewhat risky.

Conclusion

From the beginning, the Norstar project faced and met formidable chal-lenges. Northern Telecom's reputation was based on its central office switches rather than its small-business telephone systems. Their penetra-tion of the American telephone market with Vantage was low and their product distribution there was modest. And the available market window was very narrow, making speed to market essential.

Norstar was a success not only as a product, but also in terms of the NPI process. However, the Norstar success raised two new challenges for Northern Telecom. Could they transfer the lessons learned from this proj-ect to product design and development efforts elsewhere in the company? Could they introduce performance measures and mechanisms for contin-uous learning that would promote effective product design and develop-ment, even when business survival was not in question?

CHAPTER 9

THE QUEST FOR SIX SIGMA
QUALITY: MOTOROLA CORP.*

In 1989, the Paging Division of Motorola was part of the Communications Sector of this $8 billion electronics company. This Division supplied radio pagers and paging infrastructure to the global marketplace. The facility in Boynton Beach, Florida, was the headquarters for the Division, and other manufacturing plants were located in Puerto Rico, Singapore, and West Germany. Components and subassemblies for the pager products came from other Motorola locations around the world.

This chapter describes Motorola's development of the Keynote pocket pager during the period 1987 to 1989. This case history deals with many of the topics introduced in Part 1 of this book. A dominant feature of this case is the effort by an established company to design and develop a relatively low-cost, high-technology product that would reach a new frontier of quality. This case also introduces the challenge of global product design, a subject discussed more fully in Chapter 12.

COMMITMENT TO QUALITY

Confronted with an onslaught of competition from the Far East and Europe, Motorola was committed to fighting for global markets for its electronics products. One of Motorola's main strategies in this continuing battle is improved quality—in products, services, and, in fact, every phase of all jobs. In the mid-1980s, Motorola embarked on a program to improve product quality by 100 times in five years. Even as the company moved toward this goal, which seemed almost impossible when first discussed, it was realized that the result would not be good enough to remain competitive. Further, the "100X" goal spoke only to product quality and

---

*This chapter (excluding the author's added commentary section) is an edited version of the original case study prepared by Frank Lloyd (1990).

did not cover all of the other aspects of the daily jobs of Motorolans. Therefore, a new goal, designated *Six Sigma*, was formulated. Motorola defined Six Sigma as "a measure of goodness—the capability of a process to produce perfect work."

Six Sigma is measured in terms of *defects per opportunity for error* where a defect is defined as any mistake that results in customer dissatisfaction. In order for a process to be at Six Sigma it must have less than 3.4 defects per 1 million opportunities. The strength of this seemingly simple concept is that by applying the term *customer* to anyone who has a need for a good or a service, the measurement of quality can be extended to every function in the company. For example, a customer can be a manager reading a memo, the next person on the assembly line, someone buying food in a company cafeteria, an order administrator handling field orders, or even the purchase of a paging receiver. Under the rallying cry of "Six Sigma," everyone at Motorola is considered to be both a customer and a supplier.

In terms of product quality at this exalted level, it was swiftly realized that no amount of time or money could take a product designed in the past and make it reach Six Sigma. This quality level has to be designed in from the beginning. This chapter chronicles Motorola's development of the Keynote pocket pager which was intended to replace an existing product line.

ORIGINS OF THE PRODUCT

From 1976 through 1983, Motorola had introduced several families of voice pagers, each of which had models that would receive signals using one of the available encoding schemes and frequencies. It was often easier to design an entirely new unit than to retrofit an older model when a new encoding method or frequency band for message transmission appeared. Since each encoding/frequency combination required a different pager, the number of models within all of the product families had proliferated with associated problems in inventory control, manufacturing, and maintenance of each product line.

In 1987, the Paging Division decided to fund the development of a new family of voice pagers. The market, though small, was profitable, and a voice pager would ensure that Motorola fielded a complete line of products. It had been many years since a truly new voice pager was introduced, and an opportunity existed to replace the large numbers of aging pagers that had been previously sold. The demands of Six (or even five) Sigma quality had not yet been achieved in existing products.

Manufacturing and Distribution Infrastructure

The new product would benefit from the existing manufacturing and distribution infrastructure. The factory producing the existing line of voice receivers was ready and capable of handling a new product. A trained high-level distribution force was in place, selling paging receivers, so a new pager would not require incremental sales investment. Despite these facilitating factors, it was clear that the hurdles for competing in the pocket pager market were formidable. As shown in Figures 9–1, 9–2, and 9–3, paging receivers were subject to considerable technological progress through the years. Size, price, and failure rates all had sharp downward trends. Targets for the new product would have to reflect these trends plus others.

Product Strategy

In the design stage, the unnamed product was code-named Marlin, but we shall refer to it as Keynote—its final name. Keynote had certain goals

---

FIGURE 9–1
Pager Price Trends

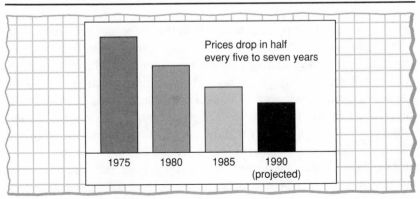

Prices drop in half every five to seven years

1975   1980   1985   1990
                        (projected)

FIGURE 9–2
Pager Quality Trends

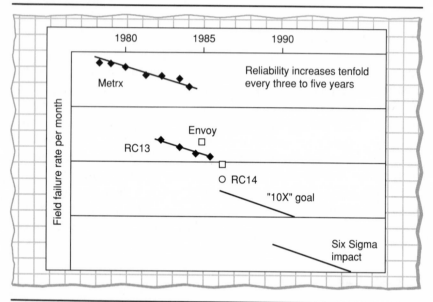

which drove the initial specification. First, Keynote was to replace every voice paging product that Motorola produced. Four different pager receivers were all to be canceled upon the introduction of Keynote. This meant that Keynote would have to function at all paging frequencies and support all of the existing encoding formats. Keynote was also to be the industry leader in terms of size, battery life, features, and styling.

Motorola's market share has traditionally been good in the domestic market. A new voice product would hold this share and even increase it through the churn expected from a smaller, lighter, more fully featured product. Two market-leading features were specified: (1) voice storage, and (2) numeric display plus voice. These features, both options to the standard product, were seen as market expanders. Voice storage, in the pager, is an answer to the problem of missed pages due to a noisy environment. The design called for storage of up to 32 seconds of digitized voice, which could be split into four 8-second messages, or two 16-second messages. Adding a numeric display to a voice pager relieves congestion on a single channel. Since 75 percent of paging messages are simple telephone numbers, these can be transmitted quickly, but important messages that require voice can still be sent to the same pager.

FIGURE 9-3
Pager Size Trends

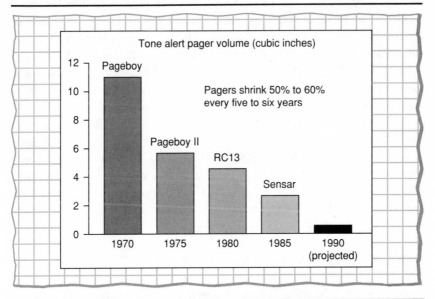

The Paging Division prioritizes programs based on a number of factors including: business need, market need, development cost, and technological risk. A new voice pager, although deemed to have market value, could not be construed to be a compelling business or market requirement for the Division. Implicit in the design was the exigency to minimize technical risk in order to reduce development time and cost.

Critical Success Factors and Priorities

Top priority in the design and development of Keynote was to achieve quality improvement to the five Sigma level. At the time of conception, no product in the Paging Division had hit this exalted quality level. In fact, a fair amount of the portfolio was below four Sigma. Keynote would be the first product to ship at the five Sigma level of quality conformance.

After quality, size had priority. Since a voice pager requires a speaker, the speaker size for Keynote had to be no larger than two inches in diameter. This posed a considerable challenge in finding a small speaker that sounded as good as the much larger speaker in the Spirit pager.

Cost was also a priority. Since Keynote had to replace many different models of existing pagers, certain savings would be accomplished simply through the reduction of the number of unique parts handled, such as efficiency of labor and larger buys based on unit volumes greater than any of the individual models replaced. However, the declining prices in the market necessitated that Keynote be lower in actual material and labor costs than the lowest of these different models. Management intentionally gave the design team an extremely ambitious cost target to achieve from the very beginning of volume production of the new product.

Beyond quality, cost, and size, Keynote required other features that the existing products lacked, such as an attached, self-locking battery door. The battery itself became an issue. While a certain group of customers preferred a rechargeable battery, others desired a commercially available "drugstore" battery that would be replaced as needed. This would be a first for a widely available voice pager.

DESIGN CONSTRAINTS AND PRIORITIES

In the initial design phase, a cross-functional team including Engineering, Marketing, and Industrial Design started the process. The small-sized parameter was established early. Motorola's smallest pager, the Envoy Tone-Alert Receiver, was chosen as the upper limit for the size. The Stored Voice option would fit the same box without modification (but the Numeric Display plus voice model turned out to require a slightly larger box).

Industrial Design led the early sessions by producing wooden block models, design drawings, and more finished detailed models. From a visual aspect, the product had to have a high-tech look. Rounded corners were designed to soften the product, giving it a modern, "upbeat" look. This approach also visually makes the product seem smaller. Various studies of different volumes were undertaken to determine if everything could fit.

The team decided that Keynote would use the same radio frequency receiver board that was already in the Bravo pager. Bravo, Motorola's existing highest sales volume pager, was designed with a separate receiver and decoder boards. (A receiver "listens" to the radio channel, while the decoder interprets the paging address and performs the requisite functions.) Since there were models of Bravo in all existing paging fre-

Courtesy of Motorola, Inc., Paging and Telepoint Systems Group.

quencies, use of the Bravo receiver provided two distinct advantages. First, the use of an existing part relieved the development group of the necessity of designing a new component. Also, building on the significant volume production of Bravo delivered certain cost savings to the new Keynote product. On the downside, the Bravo receiver placed a size constraint on the product.

The marketing specification of the decoder also called for unique changes from the past. Existing decoders for the two-tone encoding format used active filters or reeds to translate the pager address. These components are notoriously unreliable and, because of their integration with the pager, extremely expensive to replace. Customers had voiced many complaints regarding both "features." The marketing specialists on the team adamantly called for the use of a solid-state decoder to supplant these technological dinosaurs.

The critical success factors established for Keynote were based on knowledge of customer acceptance of prior products. As a voice pager, a new product should sound as good as the best product in the world. Motorola's Spirit pager was the standard by which the sound of voice pagers is judged. A relatively large unit, particularly by standards of that period, Spirit had a distinct advantage over the new design. Speakers work by moving air, and the large body allowed a larger speaker and consequently more air movement than Keynote's design. This problem was compounded by the subjectivity of what makes "good" sound.

The ability to change the identification of a paging receiver in the field, (i.e., to make the unit respond to a different address) contributed to the success of the Bravo Numeric Display pager. In other models, the unit must be returned to an authorized Motorola service shop to have its code changed. To introduce a new model that was not field programmable would be a step backward and limit sales. In the Spirit pager, the door to the battery compartment detaches from the unit when unscrewed. Even the phenomenally successful Bravo did not solve this customer complaint, replacing as it did, the round, metal battery door of Spirit with a sliding, detachable, plastic cover. Unquestionably, any new products would require a battery door that did not separate from the body when opened.

REVIEW OF PRIOR NEW PRODUCT INTRODUCTIONS

The definition team scrutinized past NPIs for relevant successful programs as well as failures. The not-so-surprising result of this activity was that the most effective NPI efforts fell into two camps. Those few NPIs that were critical to the continued existence of a group (Division or Sector) in which all available resources were devoted to the program invariably triumphed. An almost unlimited checking account helped, but of more importance was the psychological boost that the team members received from the fact that everyone in the organization recognized their importance and was behind their efforts. The second type of effective NPI were those that invested a relatively large amount of effort early in the initial design phase so that most, if not all, of the problems that appeared later in the implementation phases had already been considered.

In particular, the Bandit program, which had included the design of a new pager (actually a model of the Bravo Numeric Receiver), and a completely automated flexible production facility, was used as a model. The Bandit program thus combined the redesign of a product with the development of both equipment and software. It was implemented using a cross-functional team including design engineers, manufacturing engineers, software engineers, and suppliers. A key component of the Bandit program was the "contract book." Put together by interdisciplinary teams representing all functional teams, the contract book detailed all facets of the program. Deliverables, schedules, cost, and quality targets were delineated and signed off by the relevant managers. By fusing the efforts of all groups at the beginning of the program, Bandit achieved its goals. A similar method would be used by Keynote.

The Boynton Beach group with responsibility for Keynote was at the time deeply involved in implementing a "focused multidisciplinary team" approach to new product introduction. A general new product introduction (NPI) process flow, with all potential disconnects identified, was developed prior to the start of Keynote. The activities on the Keynote NPI were based on this work.

ORGANIZATIONAL APPROACH

Motorola organized divisions around product lines. Marketing and Development reported to the Product Manager who had responsibility for the profitability of the products under his control. Manufacturing, in the Paging Division at the time of Keynote's conception, was centralized and reported to the Division general manager. Other key support functions, such as Purchasing and Quality Control also reported outside of the Product Manager's area of authority.

The Development group, led by an Engineering Manager, worked on multiple projects simultaneously. New product creation, while of extreme importance, was only one of the responsibilities of this unit. Of equal importance was maintenance of the existing product line. This task included two major facets: (1) continuing cost reduction to maintain product margins in the face of the ever declining market prices, and (2) the monitoring of quality in terms of deviations from the accepted standards.

Manufacturing of pagers can be split into two parts. Board level assembly, commonly referred to as *front-end* processing, and final assembly, or *back-end* where the pager is given its distinctive personality in terms of individual address and radio channel. The Paging Division performs almost all of its back-end functions in Boynton Beach, with front-ends scattered around the world. One of the major front-end assembly facilities is in Singapore.

The Singapore factory handled products from a number of Communications Sector groups, including Paging. Although it reported under a totally different management team than Paging, the Paging Division had accounted for much of the efforts over the six years of its life. As a necessary component to successful manufacturing, Singapore had built up a high-level engineering group. Well-educated, energetic engineers staffed the Motorola plants.

The Florida/Singapore Partnership

At the time when Keynote was under consideration, the Paging Division had other programs under way that ranked higher in one or more of these dimensions of priority. Therefore, it was decided to split the Keynote development effort between Singapore and Boynton Beach. The single, biggest change from earlier NPI efforts was the utilization of a Singapore team for product engineering, design, and development. The Boynton team (highlighted previously) functioned in the role of expert consultants early in the program, and then took up the final factory introduction process as the pager neared completion. This presented significant challenges in communications.

Boynton Beach assumed responsibility for conceptual design, factory introduction, and some of the more intricate, technical issues, while Singapore was to do the detailed design, source parts (in Asia), run pilot lots, and initiate front-end production. Experts in speaker technology and mechanical analysis were assigned in Boynton. These people remained on the project for the duration. Additionally, Boynton had a small cadre of extremely high-level technical experts who were available on an as-needed basis at all phases of the task.

Organizational Roles

The following groups played important roles in the Keynote program:

1. **Marketing.** Provided input on customer requirements. Devel-

oped part of the contract book. In particular, defined the specifications for battery use and life, audio quality, earphone jack, flashing light for incoming message, solid-state decoder, and field programmability.

2. **Industrial design.**   Developed conceptual picture of product. Built various models. Added rounded and softened lines, with bumped out area for speaker and labels.

3. **Advanced development.**   Implemented Five Tone decoder integrated circuit, low-current EEPROM for programmable address, and assisted with the Two Tone decoding microprocessor software.

4. **Purchasing.**   Organized vendor base in Singapore—with very limited prior experience.

5. **Quality assurance.**   Established defect per unit (DPU) budget for the product. Provided extensive reliability plan. Ran all Accelerated Life Tests.

6. **Production operations.**   Derived test stations and tooling. Built prototypes. Ran front-end pilots.

7. **Product engineering/design.**   Design for manufacturing and design for automation. In particular, implemented the "clam shell" approach to mechanical assembly in which two parts can be snapped together without the use of screws or other fasteners. This design choice also made eventual robotics possible.

8. **Process engineering.**   Benchmarked current processes to identify areas where effort could be saved.

9. **Field service.**   Provided additional customer input. Defined formats for menus for field programming of pagers using personal computers.

Selecting the Program Manager

Of particular importance in staffing the Keynote program was the selection of the Program Manager. Prior to the start of this NPI, work had gone on in developing both the responsibilities of a Program Manager, and a profile of such an individual. The profile deemed that a Program Manager should possess the following characteristics:

Assertive.

Organized.

Recognized leader.

Engineering background (EE/ME/IE/Process).

Senior Engineer/Group Leader/Section Manager.

Willing to travel.

Good communicator.

Another requirement was added to the selection process; the individual had to be available. This immediately ruled out those people whose involvement with their current programs extended beyond the "drop-dead" start date for Keynote. Almost by definition, this eliminated engineers at the upper end of the grade levels delineated in the "ideal" candidate profile.

Finally senior management decided that the Program Manager had to be selected from within the group responsible for the product; that is, from the set of engineers reporting to the Product Manager whose budget would support the development and whose organization would eventually reap the rewards. Although this group was led by an Engineering Manager considered one of the best in Paging, the two most experienced Section Managers reporting to him each had an NPI of his own that, at the time, had equal priority with Keynote, and was somewhat further along the development cycle. Unfortunately, all of the other people at the next level down in the group were relatively inexperienced in program leadership.

The person picked to head the Keynote team in Boynton Beach was a Group Leader who had not previously run a major program. A highly regarded engineer, embodying some, but not all, of the character traits in the profile, it was felt that with the aid of the strong Engineering Manager, the Program Manager would grow into his role. The technical experts in Boynton Beach would ease the way.

In Singapore, the organization differs significantly from the American operations. Because the facility was founded as a manufacturing support unit for the entire Communications Sector, the traditional Product Group structure was missing, and everyone reported instead within the Manufacturing function. Therefore, the priorities, and day-to-day approach of the people, revolved around shipping product more than developing new products. Key American personnel rotated through Singapore on two-or-three-year stints in order to embed the Motorola culture and learn Asian methods. The team assigned to Keynote had little Paging experience, and the Americans heading Singapore at the time also did not all come from a Paging background.

Working with the Singapore Team

Of all of the challenges in the design and development of Keynote, none was greater than the use of a Singapore team in the product design phase. The logistics of long-distance management were compounded by the cultural differences inherent not only between an Asian culture and that of the United States, but also the dissimilarities in priorities between an organization with a manufacturing thrust and a product development group. There were also some special problems that arose independently in Singapore.

Singapore is exactly halfway around the globe from Florida. The time difference is 12 hours in summer and 13 in the standard time months. This meant that there was no time during the "normal" working day for communication. Thus the little decision points that often arose during the day defied immediate resolution. In some cases, these crises were solved on the telephone that night, which meant that people had to wait almost a whole day to restart their tasks. In the majority of cases, the situations were considered too trivial for even a phone call, so the Motorola in-house electronic-mail system was utilized. Trading these notes back and forth was almost worse than doing nothing, as often three or four messages passed, which meant three or four 12-hour periods elapsed before the issue was resolved. (Further confusing all of the communication was the language barrier. Singapore has four official languages—Chinese, Malay, Tamil, and English. The idioms used in Boynton Beach have no counterparts in the British-style English taught to the Chinese students in Singapore schools.)

One of the purposes in moving Keynote design to Singapore was to establish a nascent design center in Asia, where there was already a manufacturing facility. While close contact with production and process people greatly aids the design results, manufacturing management was measured by a set of goals that, by their very nature, distort, if not totally precluded, the product design effort.

Performance at the Motorola manufacturing plant was measured on three items: (1) shipments, (2) inventory, and (3) budget control. Emphasis was placed on keeping the existing products (the ones paying the bills) flowing to customers. In the case of a line stoppage that imperiled shipping product, all resources were marshaled to get the units back on track. There was a natural aversion to new products. These caused perturbations in the normal flow, budget problems through above normal variance, and, potentially worst of all, inventory issues. The perception of the

Keynote team was that Singapore management did not share the same commitment to the project as the Boynton development staff. In fact, this was a false perception, perhaps due to the communication barriers, as the Singapore team was deeply committed to the program.

The two reasons that the Singapore facility design center was formed were to off-load work from Boynton, and to prepare for the possibility of a global market where products would have to be designed close to the eventual customers. These motivations were not well-understood in Florida. Responding to a concern that perhaps resulted from having seen other American companies close plants after moving offshore, the U.S. staff perceived that the engineers in Asia were a threat to their jobs. This attitude showed in the reluctance of the Boynton team to embrace the offshore team. As was known all along, there was more than enough work for both facilities, and in fact, both Boynton and Singapore have grown substantially during the Keynote work.

A combination of energetic people, high-quality education, government policies, superb infrastructure, strategic location, and low costs made Singapore a high-tech haven for firms from all over the world. Every American, Japanese, and European electronics company had put up a plant on this small island within the past decade. However, Singapore's greatest natural resource, its people, was not growing as fast as the demand. This led to a relatively high turnover of talent as firms competed both for new graduates as well as experienced engineers. Motorola could not remain immune from this problem. The Keynote program more than once lost key people during critical times.

All of these complications blended to add delays to the Keynote development cycle time. In addition, management in Singapore was reluctant to send a project staff to Florida, feeling, probably correctly, that the unfriendly attitude in the United States would dampen the enthusiasm of the troops. Both staffs had difficulty working with people that they did not personally know. This extended the learning cycle of the Singapore staff, although it did give them a chance to stretch and learn from their mistakes. Florida sent an occasional person to Asia for reviews, but not often enough for direct hands-on management.

PRODUCT DESIGN AND DEVELOPMENT

This section describes highlights of the Keynote project. Topics covered include: (*a*) product design decisions, (*b*) prototypes and test cycles, (*c*) tooling and early manufacturing, and (*d*) automation.

Product Design Decisions

During the conceptual development of Keynote, Motorola initiated its corporate program in design for manufacture (DFM). An outgrowth of the corporate quality resolve, DFM became a requirement for Six Sigma products (and company training programs were developed). While intuitively, everyone involved realized that these quality levels could only be achieved by using DFM techniques, no one knew exactly how to get there. This was made even more challenging because Keynote was the first pager for which all parts were to be sourced and qualified offshore. Thus, the team had to learn as the DFM process evolved.

In order to make Keynote as manufacturable as possible, and reduce the potential number of defects embedded during the actual production process, it was decided by the mechanical engineering team leader that the entire product would be put together without screws or fasteners. In fact, not only was the pager to "snap" together, but all of the accessories, including the optional charger and field programmer would also follow this criterion. Of course, the challenge in a design of this type is not joining parts together, but keeping them together under stress. Keynote has to pass a drop test of six feet onto a concrete floor without coming asunder.

The decision to use the Bravo receiver led to further design concerns. Implicit in this decision was the fact that it could not be modified in any manner since even a minor change would negate the very reasons for using it. Along with the size and shape constraint that an existing board placed on the design was the greater problem of interfacing a voice decoder with a receiver that had not been designed with voice in mind.

Most pagers have symbols of some sort explaining the use of the various switches and buttons. As pagers increase in variety of features, these symbols have gained a degree of complexity, since the global nature of the market precludes the use of any single language. Therefore, focused studies using Motorolans from outside the Paging Division were conducted to develop a universal set of intelligible markings. Hot stamping was expensive and painted symbols unreliable. The housing itself had to have the symbols molded into it.

By law, every paging receiver must have a visible label showing information specified by the FCC (and other regulatory agencies outside of the United States). This label details the radio frequency of the unit, along with the FCC-type acceptance number, and therefore differs on each unit. Further, customers feel that private labeling adds to the value of their pagers. Each pager design includes a place for a label identifying the customer to be placed on the unit during the manufacturing process. While enormously effective in achieving Total Customer Satisfaction (and satisfying legal requirements), this process presents nightmares to the production floor. Labels often get placed in the wrong place, upside down, on the wrong unit, or missed completely. Design of Keynote included not only where to put the things, but how to resolve the biggest quality problem—the upside-down label.

During the early design phase, it became apparent that driving the size of the product to that of the existing Envoy pager would not be possible. The quality of the audio, so important in a voice receiver, required a larger pager. Additionally, the product could not achieve the specified battery life using a AAA-sized battery, so the box was expanded to accommodate a AA cell. These decisions were made very early in the program, and did not severely impact the schedule.

The initial concept for the decoder used components on both sides of the board. This allowed use of low-cost resistors and capacitors, while still maintaining product size goals. After reviewing the process, and the potential for errors, smaller, more expensive parts were used so that only one side of the board was involved.

Aesthetics and audio quality drove the speaker grill design. Unfortunately tests proved that the grill did not provide enough support for the radio. A redesigned grill, with fewer openings, was substituted.

Prototypes and Test Cycles

The method used in Keynote for prototype development was borrowed from Asian manufacturers and, at the time, was unique to the Paging Division. Historically, American manufacturers, such as Motorola, had relied on sampling plans to guarantee product quality. These samples were taken *after* the production of customer units had started. As a result, American electronics companies processed an enormous number of product/process changes during and following production, at great costs. On the other hand, Asian companies, and the Japanese in particular, ran very large prototype runs and introduced product changes far earlier in the cycle. In fact, 90 percent of the changes in a Japanese product are complete a full year before market introduction (while most American companies are just beginning to introduce changes at this stage).

Keynote manufacturing was thus developed in an iterative manner.

Six to eight cycles of building prototypes and then running large pilot lots were performed. After each lot was run, the units were put through the Motorola Accelerated Life Test, which is designed to simulate the use of a product over time, but is actually condensed into a few weeks. The policy in the Singapore factory was to evaluate and refine the product after each such prototype run. A three- or four-week turnaround time between prototype runs was possible due to the extremely close relationships established with the local parts suppliers.

The personnel used for these pilot runs were actually in training for the eventual production runs. By tracking issues and defects per unit (DPU) on every lot, a cycle of continuous improvement unfolded. In this manner, any shortcomings in the product, the production process, or the work force were all identified and solved during the prototype and test cycles.

Tooling and Early Manufacturing

Another positive attribute of Asian manufacturing is the availability of relatively cheap product tooling. The tooling for Keynote was used very early in the process. By using semi-hard tools, rather than hard tools, tweaks in the actual tools were possible between prototype runs. This allowed the pilots to use the actual tools as they were run on what was to be the final manufacturing line.

Automation

The success of the prior Bandit program had demonstrated the benefits of robotic assembly. Although there were no plans to invest in a fully robotic line for Keynote, the product was designed for potential robotic assembly in case the volume greatly exceeded forecast (or the cost of robots dropped severely). Motorola was, however, committed to selected use at the outset of certain automation technology for the volume production of Keynote. Automatic part insertion machines, necessary in a product of this size and complexity, were part of the original plan. The final test stations also used automated equipment with special tooling.

Problems in the development of the final test stations at Boynton Beach were responsible for delaying the introduction of Keynote. These stations test the radio after it has been injected with those items that differentiate it from other pagers—specifically, radio frequency and individual address. Because the Boynton factory was in the midst of an enormous volume increase in the Bravo Numeric Display receiver (and because the local manufacturing types were reluctant to commit tooling engineers to a product in early development stages), these test stations were not available until quite late in the program. In fact, they turned out to be one of the major items that would determine the starting time for shipment of the new product.

SUBSEQUENT OUTCOMES

To guarantee appropriate levels of engineering resources, and to further globalize product sourcing and manufacturing, Keynote was codeveloped in the United States and Singapore, an effort never undertaken by this Division before. This proved to be an important learning experience for the Division, committed to pursuing such global product development in the future. Another notable learning experience was the use of several prototype iterations with accelerated reliability test cycles; this approach was credited with helping to meet the five Sigma quality standards with only a few engineering changes being required in the first year.

Although it achieved all of its quality and cost targets, the first model of the Keynote series was late to market, based on the original schedule. Technical difficulties encountered in closely combining a microprocessor and a radio caused a long delay. Attempting to bring out every coding format in all radio frequencies simultaneously also caused delays. In fact, this goal was eventually relaxed and the coding formats came out over a six-month period. Sales were delayed and the older products remained in the portfolio longer than planned.

However, once on the market, Keynote met initial sales projections. Several of the new product features were hailed by customers. Shortly after the introduction of Keynote, a major competitor, NEC, announced its withdrawal from the voice pager market. Sales of Keynote were expected to rise considerably as Motorola phased out its older products.

From the standpoint of the finished product, Keynote was unquestionably a design success. Even though the best mechanical engineering minds doubted that it was possible, the pager is assembled without fasteners or screws of any type—and stays together through repeated abuse. The sliding, self-locking battery door was a big hit with the sales force, who often, in the interests of customer satisfaction, drove many miles to deliver a handful of replacement Spirit battery "buttons," wasting valuable time that should have been spent on getting new orders.

It was expected that the elimination of active filters in the two-tone model would pay dividends in the future, when more of these models would be in customer hands, so this laborious software task was considered to have been worthwhile. Finally, the low-voltage EEPROM chip used for field programming exceeded all specifications. This, too, was expected to have future ramifications as the field volume increases.

In two ways, the Keynote project team gained new skills and, in some cases, enhanced their reputations. In Singapore, a small group of people who stayed with the program for the duration had gained valuable skills in pager design and implementation in a production environment. Since they learned, somewhat painfully, by their mistakes, they probably came away with more of an appreciation for the amount of effort involved in such a project than if they had been led through the task by more experienced managers. In addition, teams on both sides of the world gained knowledge and respect for each other's problems and cultural orientation. It is very likely that in future programs these people would be able to work together better.

The Program Manager also grew during the Keynote struggle. At the beginning, his availability and position in the organization were as much of a factor in his selection as the fact that he came close to the profile of a Program Manager. Sometime in the middle of the assignment he began to demonstrate that his selection was not necessarily an error. His communication skills improved almost daily and he learned that management and leadership are not always the same thing. As with the Singapore team, this person would be an asset to the firm in future endeavors.

COMMENTARY ON THE CASE

The Keynote family of voice pagers was designed to replace all of Motorola's existing voice pager products. It was intended to be the industry leader in all features including: cost, sensitivity to the radio signal, small sized, lightweight, long battery life, ease-of-use, and quality of the sound. This new paging receiver had to function within the existing set of governmental constraints and paging infrastructure. Consistent with corporatewide goals Keynote was designed to meet five Sigma quality levels, the highest standard ever imposed on this kind of product.

In contrast with NeXT, a start-up company which approached product design and development with a clean sheet of paper, the Motorola Paging Division began the Keynote project with what could be called *a dirty sheet of paper*. This image is central to appreciating the lessons from the Motorola case history because the dirty sheet of paper brings both enablers and constraints to the process of product design and development.

The foremost enabler was probably the existing corporate commitment to quality and the five Sigma standard shared by all who participated in the Keynote project. Also important was the device of the contract book through which all key participants would commit, in the planning phase, to the objectives, specifications, timing, and responsibilities associated with the project. Another important enabler was the existing manufacturing and distribution capability.

Constraints derived primarily from ongoing business and other projects were: the need to support the existing product line of voice pagers; and the requirements of other simultaneous new product initiatives by the Division. As stated in the case, the design and development of Keynote was, relatively speaking, not a high priority. Whereas the Norstar project at Northern Telecom benefited from the crisis facing the Division making and selling small-business telephone systems at that time, Motorola did not feel it necessary to mobilize a comparably extraordinary effort around Keynote.

Structuring the Work

From early design through prototype development and test, the Keynote project illustrated important aspects of effective NPI problem solving. At the concept development phase, industrial designers were instrumental in helping the team decide on the size and shape of the Keynote receiver. The product was successfully designed to be assembled without the use of fasteners or screws of any type. The Keynote project also illustrated the importance of linking prototype development with test cycles and the use of semi-hard tooling to facilitate design changes between prototype runs. Another valuable practice was the design of the product for eventual assembly by automated equipment even when the initial plan was for manual production.

Planning and Managing Projects

Keynote had clear quality objectives which dominated concerns for development cycle time or development costs. This dominance of specific aspects of product quality, coupled with the existing corporate culture that supported the quest for extraordinary levels of quality, provided a common focus for the various project participants. Under this supportive condition, the lack of an experienced project manager was less of a serious liability than it might have been. Demanding targets for the size of Keynote and its unit cost were also important and needed to be balanced with concerns for quality. To reduce the development cost and risk, the team decided to use proven modules from existing paging receivers, thus placing additional constraints on the design process. Keynote's Program Manager did not have ongoing authority over the many participants in product design and development who were permanently assigned to functional groups that served the needs of both existing and new products.

Developing Collaborative High-Performance Teams

Keynote was an example of global new product development, of a type that the Motorola Paging Division had not previously experienced. Staff from the Singapore plant were responsible for product engineering, design, and development and front-end manufacturing introduction, while Boynton Beach had responsibilities for conceptual design, factory introduction and back-end manufacturing introduction. All parts were sourced and qualified offshore. This project offered Motorola needed learning opportunities in the handling of such globally dispersed responsibilities. It is important to note that many companies are faced with similar global opportunities. In such cases, the notion of fully co-located product design and development teams is simply not an option. The complexities and delays that Motorola experienced are probably typical of what to expect, even when the challenges are reasonably well-understood. Motorola could also build on their Keynote experience in strengthening this aspect of future design and development projects.

Coping with Technology Choice and Risk

Because Motorola wanted to minimize the time and cost to introduce Keynote, they deliberately limited the extent of new technology in this product, subject to perceived customer requirements. Accordingly, they decided to use the same radio frequency receiver board that was already in use in one of their existing pagers. Marketing specifications, however, called for new development activity to improve the reliability of the decoder board. To simplify manufacturing, a snap-together design (no screws or fasteners) was adopted, and the extent of initial use of automation was limited.

Conclusion

Motorola was fortunate to be in a position where time-to-market was not the critical success factor. They ended up needing more time than planned due in part to the special organizational complexities of global design and manufacture. No compromises in their extremely high-quality standards were made. One gets a sense from the Keynote case history that Motorola had already begun to view product design and development as an area ripe for organizational learning. This project was seen as a stepping-stone along the path of continuous improvement of a vital business process. Motorola's philosophy of Six Sigma quality management seems to have moved seamlessly from the field of manufacturing to the process by which new products are introduced.

CHAPTER 10

DEALING WITH SIZE AND COMPLEXITY IN A REGULATED ENVIRONMENT: GE AIRCRAFT ENGINES*

General Electric (GE) is among the largest U.S. manufacturers, and their aircraft engines business, in 1983, had net earnings of $196 million on sales of $3.5 billion. Research and development spending was $600 million, half of which was government funded. In that year, GE Aircraft Engines (GEAE) had 28 percent of the world market for all engine types, including 30 percent of the world commercial jet engine market.

In 1965, GE was awarded an Air Force contract to develop the engine for the first supersized transport (Lockheed C-5A Galaxy). The engine that resulted (TF39) marked the introduction of the high-bypass turbofan. General Electric subsequently adapted this technology for use in a series of commercial jet engines in the CF6 family. After unsuccessfully proposing that United Airlines purchase numerous CF6 engine derivatives for the Boeing 767, a decision was made to develop a new engine, the CF6-80A. This product development program was launched in late 1978, when commitments were made by American and Delta Airlines.

In 1978, the market for a new engine was driven by an increase in air travel, and airline operating profits were rebounding from the recession levels of the early to mid-70s. Fuel costs were projected to reach $2 per gallon, which generated a need for more fuel-efficient engines. With applications to the new Boeing 767 and Airbus A310 wide-bodied aircraft, the CF6-80A was GE's hope for sustaining the commercial market penetration that had taken place since the introduction of the CF6 in the early 70s.

The development of a new engine has a significant development cost

---

*This chapter (excluding the author's added commentary section) is an edited version of the original case study prepared by Donald Gregory (1990).

(approximately $1 billion), complex product technology, and very long lead times, which were typically around seven years in the 1970s. Because of the costs and time, the engine manufacturer must constantly try to anticipate changing requirements and to try to develop advanced technology that could be used in the development of future aircraft engines.

This chapter describes GEAE's design and development of the CF6-80A commercial jet engine during the period from 1978 to 1982. This case history illustrates how a most complex product was introduced by a large number of people over a significant period of time, in a regulated environment. The case is further characterized by a firm and ambitious product development cycle time and a major subsequent cost reduction effort.

Courtesy of GE Aircraft Engines.

DESIGN CONSTRAINTS AND PRIORITIES

The typical engine development schedule, as shown below, was seven years (see Figure 10–1). The introduction of the CF6-80A for use in the Boeing 767 required a shortened development cycle time to four years. This time-to-market goal became a principal driver in the introduction of the CF6-80A.

The objectives of the CF6-80A project, in priority order, were to meet specific targets for product performance, life cycle cost, and manufacturing cost. Meeting these objectives required a number of design changes to basic concepts developed in earlier commercial and military applications, such as improved fan blade and rotor designs, a new placement for the gearbox, and elimination of the turbine midframe. Extensive use of outside technology—from their many vendors as well as an international partner—was required. Outsourcing, as was typical, accounted for about 65 percent of total product cost. Through considerable interaction with the airframe manufacturers, a balance was achieved between engine design/technology requirements and airframe designs/performance.

FIGURE 10–1
Typical Engine Development Schedule

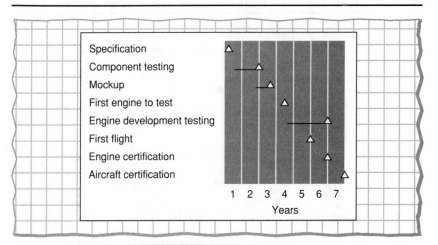

The Design Strategy

The key in the new product introduction (NPI) decision process is the design strategy for the new engine itself. This design strategy is based on whether a current product or a product already in the development cycle can be used as a technology base and the new product can be a derivative of the base product. Each major engine component (e.g., fan, compressor, high-pressure and low-pressure turbine—see Figure 10–2) has its own limitations for performance upgrading.

The CF6-80A was a derivative of the CF6-50 and as such the design, development, manufacturing, and certification effort was less than a completely new design. The CF6-80A project required technical innovations upon existing engines to provide a lighter, cheaper, and lower fuel consumption; shorter length; improved fan blade containment; and a core-mounted gearbox.

Product quality—which is a given requirement without any compromises—would require improvements that would reduce the shop visit rate, reduce in-flight shutdowns, and improve dispatch reliability. These factors were relative to the CF6-50 which, at the time, was establishing

FIGURE 10–2
Schematic Diagram of a Jet Engine

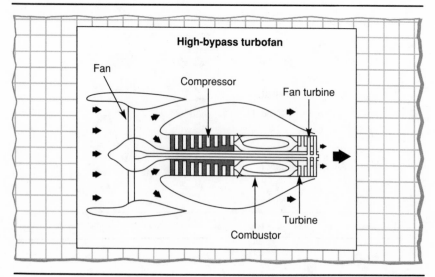

industry-leading standards in product quality. Other factors were the profit potential that would be generated by the expected demand and the CF6-80A with both 767 and A310 applications.

The management of manufacturing costs was based on learning curve concepts. The conceptual design was costed at the 250th unit by the Advanced Value Engineering Organization. Although this cost estimate did not involve the Value Process Engineer in the production shop, the personnel involved have significant experience in the manufacturing process. The production shops projected an updated 250th unit cost at the sign-off of the part drawings. Learning curves were used based on historical data, pre-load (experience of similar products), and schedules derived from the operating plan and/or business plan in a longer time frame. Purchasing provided the raw material costs and Finance provided overhead rates, hourly rates, and escalation factors. The design itself did not employ a mature first-cost philosophy. Periodic cost targets were continually updated based on the changing environment (primarily schedule forecasts) and in light of the cost objectives not being met.

Aircraft Mission and Operational Requirements

The design process is driven by the aircraft mission requirements, which specify how the aircraft is expected to perform in flight. In addition to mission requirements, there are also operational requirements that define the desired maintainability, logistic, and serviceability characteristics of the aircraft. Both mission and operational requirements are specified prior to the design of the aircraft, are independent of the design, and are formulated by the end user or customer. Multiple mission and operational requirements invariably lead to conflicting design choices. This usually occurs during the conceptual and preliminary design phases when the realism reflected in the requirements is still being assessed. Consequently, mission requirements may often be reappraised and compromised as the design of the airframe and propulsion system progresses.

Figures of Merit

Figures of merit (FOM), on the other hand, are performance and operational characteristics dependent on the specifics of the aircraft design, but are not directly specified in the requirements. The designer use the FOMs to evaluate competing designs which satisfy the requirements. In some

FIGURE 10–3

Typical Mission Requirements and Figures of Merit for Commercial Transport

Design Requirements	Figures of Merit (FOM)
Range	Initial investment
Payload	Direct operating cost
Balanced field length	Cost per seat mile
End of climb thrust	Fuel consumption
Engine out climb gradient	Fuel per seat mile
Noise and emissions	
Growth capability	

cases, the distinction between an FOM and a requirement may not be obvious, especially if there is little or no operational experience associated with the system. Generally speaking, however, the mission and operational requirements drive the design options while the FOMs drive the selection from within these options. Formulation and recognition of the relative importance of the proper FOMs is critical to a successful design effort and requires a considerable understanding of the user's needs. Attention to the wrong FOMs can obviate an otherwise successful design.

In general, commercial systems will be evaluated primarily on economic factors; although environmental and safety constraints also receive considerable attention. Fuel costs in some form will be the principal FOM. However, purchase price, ease of maintenance, reliability, direct operating cost (DOC), life cycle cost (LCC), take-off weight, and so forth, all play an intertwining role in the design of the airframe and propulsion system.

Typical commercial transport mission requirements and FOMs are shown in Figure 10–3.

Program Schedule

The CF6-80A program was started in August 1978. The original development program schedule is depicted in Figure 10–4. The schedule was very aggressive and targeted the first engine to test (FETT) in November 1979. This represented a cycle time reduction of nearly 27 months from the typical FETT cycle of 42 months. The schedule set the tone for the

FIGURE 10–4
The CF6-80A Development Schedule

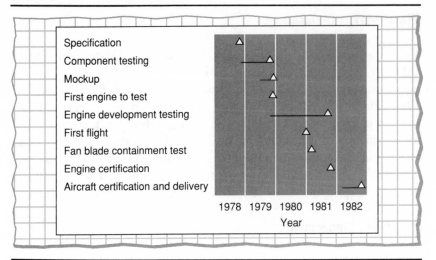

Specification
Component testing
Mockup
First engine to test
Engine development testing
First flight
Fan blade containment test
Engine certification
Aircraft certification and delivery

1978 1979 1980 1981 1982
Year

overall program management process and had a major impact on the decision making during the NPI.

This complex project was carefully managed with the development cycle time target in mind. Decisions were constantly made favoring product availability over cost. Manufacturability was not addressed initially.

Overall, the actual completion of major milestones and engine deliveries met and/or exceeded this very ambitious schedule. In October 1981, the CF6-80A engine received FAA certification, less than two years after the first engine went to test. The receipt of the certificate culminated the primary design, development, and certification efforts.

ORGANIZATIONAL APPROACH

GEAE has traditionally been organized in a matrix management fashion. Product responsibilities are managed by a small and tightly integrated project organization. Each separate product project is supported by divisionwide functional organizations (Manufacturing, Engineering, Product Support, Marketing, and Finance) that assure the most effective use of personnel and facilities. The CF6-80A project was organized in the same manner.

The Program Department

The CF6-80A Program Department reported to the Commercial Engine Division which reported to the GEAE Group Executive. The organizational structure remained the same during the NPI except for a change in 1982 which created two Production Divisions, one in Evendale, Ohio, and one in Lynn, Massachusetts. The most significant difference was that the Production organization consisted of both engineering and manufacturing under one manager. The merging of these two key functions was intended to encourage more communication and understanding between the two groups at all stages of production and thereby reduce costs.

Program Management

The CF6-80A program management process was headed by a General Manager whose experience included extensive engineering and project assignments. The General Manager reported to the Vice President and General Manager of the Commercial Engine Programs Division. The key responsibilities included setting the overall program objectives, developing specifications based upon the customer requirements, approval of design changes, and overall program management.

The Work Breakdown Structure

The overall program was managed using a work breakdown structure (WBS). There are general activities such as program coordination, cycle and performance, materials and processes, and aircraft integration. For each engine component improvement, the WBS is broken down to include specific design, manufacturing, and test activities. The WBS will vary depending on the engine component. For a single component, typical WBS activities might include:

- Aero design.
- Mechanical design.
- Materials definition.
- Materials procurement.

- Component fab.
- Bench test.
- Assembly and instrumentation.
- Development test/ endurance test.
- Data analysis.
- Results and conclusions.
- Certification tests.
- Release to production.

The Design Subteams

The CF6-80A design effort was comprised of 22 disciplines and responsibilities. Some of the subteams had functional responsibilities such as: specifications; manufacturing; evaluation; systems; vibration analysis; and support. Other teams were responsible for particular components of the engine such as:

- Core redesign and overall mechanical.
- Mounting and technical analysis.
- Low-pressure turbine (LPT) aerodynamics.
- LPT frame.
- LPT rotor.
- LPT stator.
- Compressor rear frame.
- Combustor.
- Compressor.
- Fan.
- High-pressure turbine (HPT).
- Lube system and aft sump.

Each design team established a program plan, including specific tasks/activities and levels of effort. Typically, major milestones depicted material releases in three months and design releases in six months. In total, hundreds of people were involved in this new product introduction over a period of four to five years.

The Program Control Board

One of the key NPI integration mechanisms is the use of a Program Control Board. The Board, headed by the Program General Manager, monitors and controls program changes prior to expenditure of funds to evaluate trade-off options to ensure optimum integration and provides operational direction on program changes to functional contributors and project personnel. The Program Control Board provides a means for:

> Orderly and expeditious assessment of the need for action expected to result in program and/or product design changes.

> Continuous monitoring of progress and approach, including applying technical and business considerations in resolving problems that arise.

This Board is also made up of permanent members from Engineering, Manufacturing, Quality, Finance, and Airline Support Engineering. Other members, at the discretion of the Chairman, can include production program and technical requirements managers, and field service personnel.

THE PRODUCT DESIGN/DEVELOPMENT PROCESS

Several options exist for converting engine power to propulsive power, and the most appropriate configuration is determined from the particular mission for which the aircraft is being designed and from the thermodynamic factors that control the power and efficiency of the engine. Also, there are a number of opportunities for the engine and airframe company to interact and evaluate the system design and performance.

The Stages of Aircraft Engine Design

The design process is usually divided into three levels of design: conceptual, preliminary, and detailed designs. There is a certain degree of interaction between the levels of design and, consequently, the boundaries are not always distinct. There is no absolute procedure for the design of an aircraft engine; the steps involved will depend on the organizational approach, as well as the specifics of the application. A new engine will always require more analysis than the modification of an existing product.

Conceptual Design. Mission objectives and requirements are defined. Parametric trade studies are conducted by both the airframer and engine manufacturer using preliminary estimates of aerodynamics, engine component characteristics, weights, and so forth, in order to establish the best engine/airframe match for the given mission. More than one engine-type may be considered. Cycle selection and engine sizing studies are conducted to define the engine airflow capacity, bypass ratio, overall engine pressure ratio, and other independent design variables which maximize the figures of merit upon which the system is evaluated while simultaneously meeting the mission requirements. Because of the strong interaction between engine and aircraft design variables, considerable interaction and iteration between the airframer and engine company is generally required. The process is one of compromise.

Preliminary Design. The configuration established in the conceptual design phase is refined. More sophisticated procedures are used to optimize the design variables and minor adjustments are made in order to minimize the impact of any single variable on the figures of merit. Better estimates of component characteristics are incorporated into the cycle model and greater attention is given to secondary effects. Design tools are becoming more complex and increasing emphasis is placed on maintaining the aerodynamic and mechanical integrity of the design. Manufacturing, cost, and maintenance considerations are incorporated. The configuration is beginning to get locked in.

Detailed Design. The configuration is "frozen." Detailed aerodynamic and mechanical design is completed. Accessories, controls, customer interfaces, and other components are designed. All equipment and hardware items are specified and designed. Detailed drawings are made. Tooling and other production fixtures are made.

Product Design Decisions

The CF6-80A design activities were officially kicked off in August 1978. The design itself was not started with a blank sheet of paper. Product and/ or component design concepts were being analyzed and evaluated prior to the August launch in anticipation of a new product need and performance improvements planned for the existing CF6 engine family. Specific fuel consumption (SFC) performance improvement road maps had been developed.

In order to meet the required technical innovations for improved performance, various design alternatives were evaluated. These design parameters were then translated into more detailed designs and alternatives. Whenever there seemed to be a design opportunity that would result in weight reduction, confidence and risk factors were assigned. This resulted in over 60 major design changes at the engine component level (i.e., fan, booster, compressor, high-pressure turbine, low-pressure turbine, and nacelle). In all cases, the expected improvement potential was greater than quoted allowing for margins in the design. Each item or component improvement constituted a small portion of the aggregate improvement being sought.

The major design risk involved eliminating the turbine midframe (TMF). This, along with the shortening of the combustor, would achieve the requirement for the overall reduction in engine length of 18 inches. Elimination of the TMF also reduced weight cost and the need for the TMF liner which was a high-maintenance item. On the downside, this design alternative required major innovations in the design of the sumps, seals, and bearings. Engine vibration and how to support the turbine were a major concern. Other risks involved a new fan blade containment (use of Kevlar), core-mounted gearbox, and meeting the overall SFC improvement.

A major review of the CF6-80A design concepts was held at the end of August. This review involved the very senior division and group level executives, each one of the executives having a design engineering background and many years of jet engine design experience.

In the early planning of these activities, an FETT milestone of July 1980 was anticipated. As it turned out, a faster than planned NPI cycle evolved and became one of the key objectives.

Based upon the objective to meet the First Engine to Test in 12 months, the procurement and manufacturing cycle had to be significantly reduced. As a result, the emphasis put on performance (thrust, SFC, weight) and schedule dominated costs as part of the decision-making criteria.

The engineering design was the principal driver, and innovations were drawn from the technology pool and from potential alternate designs with no focus on cost or manufacturability. The design in the form of material envelope definition was "thrown over the wall" to manufacturing to procure the needed hardware. Manufacturing personnel, given the challenging short cycle and time pressures, reacted accordingly.

MANUFACTURING DECISIONS

The GEAE manufacturing environment can be described as low-volume production, an everchanging product mix, and sophisticated processing of very high-value component parts that are made from exotic materials. The product demands unforgiving product quality and reliability.

Manufacturing Sites and Sourcing

There are 10 GEAE component manufacturing sites, comprised of satellite facilities or focused shops within the larger facilities throughout the United States. The manufacturing requirements are also supported by coproducers on a worldwide basis and a huge vendor base which contributes 65 percent of the manufacturing cost. Manufacturing is also supported by a development shop for producing prototype and development hardware.

The capabilities of these component shops play a major role in the make/buy and source selection decisions, which are dependent on previously established charter sources for certain components. Any change to these sources are reviewed by a Make/Buy Board which is comprised of senior level personnel representing the functional organizations. The decisions are based upon an analysis that includes economics, capabilities, such as capacity, proprietary processes, and source substantiation. The make/buy structure is continually challenged based on changing conditions and cost reduction programs. Changes of the sources in some cases may require requalification, which could involve engine tests. Engine tests and retooling can be most expensive, which dampens and restrains from making many changes. Because of this, it is preferable to introduce new products in the production source rather than using the development shop and possibly re-qualifying the process and/or retooling.

The CF6-80A Manufacturing Process

The CF6-80A NPI Manufacturing Process was driven by the schedule to get hardware to meet the short first engine to test and engine certification cycles. For most of the component parts, the vendor and hard tooling lead times were limiting. Alternatives available to the value/process engineer to reduce lead times was to use hardware from other CF6 programs or purchase material in bulk sizes which the dimensional envelope of the

CF6-80A would fit. If new orders were placed on the raw material vendors, deliveries would have to be negotiated to meet the schedule requirements and more than likely would have an associated premium cost.

Production of the CF6-80A NPI was assigned to existing facilities and utilized existing process capabilities in most cases. Prior to the transition to production, the internal sources component parts were either manufactured in the development manufacturing shop or the production component shops whose charter fit the CF6-80A components.

Special Roles of Manufacturing Personnel

Within each component shop (e.g., the rotating parts shop in Evendale), the Value/Process Engineer (VPE) was responsible for all aspects of the part manufacturing and its processing. The planning and implementation of the processes involved use of the Design Release Bulletin which provides preliminary descriptions of all parts. The VPE worked with Production Control to establish capacity availability. The VPE became involved when the final drawing was being detailed on the drafting board and/or CAD (computer-aided design) terminal. The VPE was familiar with similar processes from earlier designs (including 15 years' experience with similar configurations, spooled rotors, and long fan shafts) and most recent CF6-50 learning. The short cycle prevented use of hard tooling/fixturing and shaped material configuration (raw material definition).

While the CF6-80A did not require specific new skills in the work force, the planning and execution of the work force deployment involved:

1. Use of the long-range business plan to determine incremental facility and manpower requirements (this was reviewed on issuance of any new and/or updated plan).
2. Establishment of new plants to increase and duplicate existing capabilities. This required training at the new sites and the transfer of management from the skilled locations.
3. Assignment of a specialist to each production shop to determine facility/investment/manpower requirements. These specialists support a central manufacturing engineering (ME) staff with information.
4. Support of the production shops and central ME staff with computer-aided resource planning tools and use data supplied by the VPE and/or data from the electronic production routings.

An independent quality organization (reporting into the Group VP) developed the quality plans in conjunction with the production shop VPE. The CF6-80A quality metrics were the same as previous programs.

Manufacturing plans were developed from engineering drawings and specifications and were dependent on process attributes. For example, for the N/C (numerical control) processes the first piece was inspected as well as those designated from a sampling plan. The inspection steps were at independent stations between operations or on the machines. There was 100 percent final inspection for the clearance and completeness of the paperwork and operation completeness. The quality plans were very detailed and specific. They included sketches, dimensional, and other characteristics of the part, and processing steps. Training for new employees or those unfamiliar with the process was the responsibility of the Quality Control Engineer and VPE.

Initial material releases for seven sets of development hardware were released in December 1978, four months after program start. An additional 19 sets of production hardware were released in March 1979. Because of timing and the lack of available hard tooling, some development sources rather than the production shops were utilized to produce parts. The next block of production engine material releases for 170 sets were released in November 1979.

Vendor Capabilities

During the initial CF6-80A NPI time period in 1978, the entire aerospace industry was operating well above capacity and booming. This economic climate had the usual implications for lead time planning. Because of the heightened demand, lead times were expanding dramatically and early release of requirements became necessary to lock up vendor capacity. Other alternatives which were being pursued were to release sketches of CF6-50 raw material drawings and adding extra stock to ensure that the envelope would be sufficient to be able to machine the CF6-80A parts. As well as the lead times becoming greater, the costs of the base raw materials were escalating at annual rates up to 100 percent and during the period from September 1978 to September 1981, the price of some titanium, Inco 718, and Cobalt alloys had escalated 60 to 160 percent.

The limited capacity within the industry introduced the need to find alternate sources. This required extensive evaluations and technical support to qualify these new sources. In some cases, it was prudent to establish dual sources to ensure deliveries to meet the compressed schedules.

SUBSEQUENT OUTCOMES

Just as the marketplace had taken off dramatically in the late 70s with accompanying industry capacity strains, the bottom fell out in 1980. Airline profits were squeezed by deregulation, causing the introduction of new airlines, eroding fare structures, and changing route structures. At the same time, the recession produced significantly lower air travel and high-interest rates. These events produced lower airline operating profit and caused the airlines to slash operating costs, delay and/or cancel aircraft orders, and switch to the use of smaller aircraft. Lower fuel prices than anticipated also led to lower demand for the new aircraft. These factors contributed to lower CF6-80A volume, as well as significant reductions in the rest of the commercial engine base forecast load.

After successfully meeting the very aggressive First Engine to Test and Engine Certification schedule commitments, GEAE was faced with two major problems: (1) the adverse situation in the marketplace and the industry; and (2) the need for a comprehensive effort to reduce the unit product cost.

As a result of the emphasis to meet the schedule requirement and the limited capacity within the industry, the program was faced with a number of negative cost impacts. The overrun of the cost projections had multiple sources, primarily including:

Use of "soft" tooling versus long lead hard tooling.

Long lead exotic material costs.

Use of material blanks (hog-outs) versus castings (shape and size) to part definition.

Use of CF6-50 forgings (met material specs) versus forgings made to meet CF6-80A part definition

Use of larger forging envelopes to offset potential design changes to meet performance. Because of industry conditions, approximately 170 engines were released two years prior to engine certification.

Material costs were also higher than expected because of a sellers' market (exotic material escalation), buying place in line to meet short cycles, establish new and/or dual sources to protect schedule.

Higher internal production costs were a result of using many new development sources versus development learning at production sources.

The use of hard tooling missed early production because of long lead time and the producibility issues were not addressed early in the design.

The cost reduction activity itself was late getting started because of the shortened NPI cycle and the engineering and manufacturing resources were overextended to assure meeting the certification requirements.

The Cost Reduction Program

A major CF6-80A cost reduction program was launched in August 1981. This program was organized and chaired by a representative from Engineering; other representatives on the cost coordination committee were from Manufacturing, Materials (Purchasing), Value Engineering, Quality Control, and Metallurgy. The committee's main responsibilities were to manage the overall program and to facilitate the cost reduction teams. Sixteen teams were organized by engine subassembly or system and were comprised of approximately 300 members from the Engineering, Manufacturing, Materials, and Quality Control organizations. Teams were also established for major generic cost savings opportunities including the areas of forgings, structures, processes, rolled and welded rings, surface finish, fasteners, and inspections.

As the cost reduction activity got under way, it became evident that it would require a growing number of design changes to meet the projected 250th unit cost. It was a practice to provide engine cost estimates, and forecasts were traditionally based upon the 250th unit cost. Learning curves were then used to project costs in the ongoing production years for the various components, based upon part commonality and historical production. Cost reductions were sought from all sources. The net savings were then calculated recognizing the additional costs that would be incurred along the way, such as engineering cost from design changes, development hardware cost from re-qualification, additional tooling and termination costs. The impact of all changes were monitored on an engine by engine basis. Major all-day division level team reviews were held monthly to review the progress and help to solve any challenges that would surface.

Unfortunately, at this time, marketplace changes and industry conditions (previously described) were already adversely impacting the cost reduction strategy. Once again, projected shop costs were exceeding expectations. This led the Group Executive to establish a CF6-80 cost task force composed of experienced senior level members representing all the functional areas. These members defined specific objectives, data, and tasks for key individuals from Manufacturing, Materials (Purchasing), Finance, Engineering, and Project. The tasks included businesswide implications and multiple engine programs, not just the CF6-80A. These assignments set in place specific functional and cross-functional examination of methodologies, rationale, and practices.

The results of the cost reduction task force were presented to the Group Executive and top management in April 1982. Some of the major conclusions and recommendations were:

1. Engineering, Manufacturing, and Finance establish parts cost targets by source. No drawing released over the cost target.

2. Project, Engineering, and Manufacturing can reassign the individual parts cost targets, while maintaining the aggregate targets (trade-offs of weight, specific fuel consumption (SFC), and costs). Project management has overriding conflict resolution authority.

3. Review value/manufacturing engineering methodologies for achieving best process and lowest cost.

4. Analyze the manufacturing operation for each part from start to completion (eliminate, combine, improve process stability and repeatability).

5. Change in responsibilities:
   a. Shop managers responsible for farm-out costs.
   b. Division level Manufacturing Review Board for major material buys, tooling costs, facilities.

6. Bring forth the meaning of ownership and responsibility.

7. Implement detailed plans (road maps) to drive down material and overhead costs.

8. Continue high-level management cost reviews on monthly/quarterly basis.

9. Increase utilization of mechanized cost reporting.

10. Increase utilization of computerized shop floor control systems.

11. Measure component shops on individual overhead rates.

12. Continue to upgrade front-line supervision.

13. Change historical buying practices which promote complacency; balance short- and long-term objectives.
14. Manufacturing program organization should spend more time on cost reduction plans.
15. GEAE has many outstanding people.

This cost reduction activity had long-range implications and was a positive learning experience for GEAE management. The task force had a major impact not only on the CF6-80A program, but also on the entire large engine base. The alternatives that were evaluated and developed proved to be most valuable in future engine product introductions, as well as the ongoing programs.

Market Impacts

The CF6-80A NPI was a technological success. Changing market conditions and aircraft sizing had a significant impact on the expected production volume of the CF6-80A. Because of the lower volume, the CF6-80A program was a financial loser in a tactical sense. From a strategic view, the CF6-80A program was very successful in that it kept GEAE in the commercial engine market and was used in the introduction of the 767-200 and A310. The program also enabled GEAE to use the CF6-80A as a basic design for growth into the CF6-80C, which has been a very successful program and enabled GEAE to become the commercial engine leader. The commercial engine market leadership changed from a Pratt & Whitney ratio of 7 to 1 in 1983 to a GEAE ratio of 1 to 3 in 1987. Some of the reasons for the leadership change have been better customer service; GEAE becoming the technological leader; close working relationships with the commercial engine customers (including both the airframe manufacturers and the airlines); strategic alliances with SNECMA (French), and CFM-International; and an aggressive approach to the international market to build global market share.

The technological success of the CF6-80A can be measured by the specific fuel consumption (SFC), which is 6 percent better than the CF6-50C/E and met the established design criteria. The CF6-80A has superior product reliability as measured by its dispatch reliability, inflight engine-caused shutdown rate, and shop visit rate. The CF6-80A was also the first engine to be certified with the FAA's ETOP (extended twin operations) approval process (for extended range operations). The ETOP certification has made a significant impact in the airline industry, allowing more flexibility in the routings of twin-engine aircraft and associated operating (fuel) economics. The CF6-80A passed 1 million hours of flight time by 1986 and was recognized for its outstanding in-flight record. A 1986 Boeing report "credited the -80 with setting a new standard of reliability for the large turbofan engine," said Lee Kapor, Vice-President of GE's Commercial Engine Projects Division. Experience through August 1990 demonstrates continued outstanding reliability. The CF6-80A also has lower direct maintenance costs because of its rugged steel compressor, long-life variable stator bushings, and long-life cost effective combustor. All of the measurements equaled or bettered the NPI design criteria.

The CF6-80A engine has been in service on the 767 since October 1982, and on the A310 since March 1983. By early 1990, this engine had logged a total of more than 4 million flight-hours, while maintaining the tradition of the CF6 engine family for high performance and reliability. The engine's current dispatch reliability rate of 99.94 percent translates to fewer than one CF6-80A-powered aircraft departure per one thousand is delayed or canceled due to engine problems. Fewer than one engine-caused in-flight shutdown per 300,000 flight hours makes the -80A consistently superior to most other high-bypass turbofan engines. This engine has been selected by 17 customers worldwide to power more than 160 aircraft.

Changes to the NPI Process

The CF6-80A provided a basis for subsequent product enhancements that were embodied in the CF6-80C engine, which entered revenue service in October 1985.

Based on the experience with the CF6-80A engine introduction, a series of significant changes were made to GEAE's process of new product introduction. These changes, implemented for the subsequent CF6-80C project and thereafter, included organizational structure, personnel capabilities, and NPI procedures. Some of the major infrastructure changes were:

- Teaming (Design Engineering and Manufacturing) to create a mutual understanding and appreciation of functions.

- Development of producibility trade-offs over a wide span of procured and internally manufacturing components. This knowledge has been promulgated over a large GEAE population.
- Development of cost evaluation methodology to include all variable and fixed cost factors, cost/time relationship, and measurement program.
- Evolution into refined design-to-cost, multifunctional practices.
- Establishment of formal procedures.

Two organizational changes bolstered the overall NPI efforts. The first change involved combining product engineering and manufacturing. The second organization change established Product Design and Operations Control Department with responsibilities for:

- Engine systems engineering.
- Engine evaluation.
- Configuration control.
- Engine production programs.
- Schedule and cost.

Each step enhanced the relationship of Design Engineering and Manufacturing for improving the integration of the two functions.

More emphasis was also put on the preproduction planning functions, which is the responsibility of the Manufacturing Program Management. Since then, Manufacturing has had more of a voice in the NPI process as well as participating in the decision-making process. Prior to the CF6-80A NPI, this organization was in an expediting role and messenger of primarily "bad news about schedule and costs."

New technology such as CAE, CAD, and CAM (computer-aided engineering, computer-aided design, and computer-aided manufacturing, respectively) did not play a large role in the design and development of the CF6-80A. However, it has impacted more recent new products by providing capabilities for improved analysis, simulations, and control. Without changes to the infrastructure, the new technologies would not be as effective.

In summary, the CF6-80A was introduced under extreme time pressures and during a vulnerable industry time period. Despite these conditions, the strategic value and the outstanding product performance of the NPI has proven to be very beneficial to GEAE.

COMMENTARY ON THE CASE

A jet engine is an intricate product that can only be designed and developed by large numbers of people with diverse and highly specialized skills. It is also a product with extreme safety implications and, accordingly, its development is extensively regulated by government. Of the four case histories presented in Part 2, this one is intended to illustrate the challenges of product design and development under conditions of size and complexity. Admittedly, relatively few people ever get to manage (or participate in) the design and development of a jet engine. However, as with the other cases in Part 2, the reader is encouraged to glean from this case lessons that are transferable elsewhere.

Structuring the Work

The GEAE case contains extensive descriptions of the way that the work of product development is structured and managed. This topic alone could form the basis for a book on managing complex technology projects. The reader cannot help but notice that the terminology is both extensive and heavy, reflecting considerable formality: work breakdown structures (WBS), Program Control Board, Make/Buy Board. Although this approach is different from the phase-gate management structure described in Chapter 2, it has the same objective—to provide the necessary frequency and types of management control.

At GEAE, as with other companies we have examined, considerable work precedes the start of the full new product program. Much of this work is of a planning nature, along the directions described in Chapter 2. GEAE doesn't use a single contract book, as Motorola does. Instead they have an elaborate system, suited to their product and their organizational structure, of allocating time and money constraints, coupled with component cost, weight, and other performance attributes to many different project subteams. GEAE projects are so complex that they have subteams devoted to providing specialized forms of staff support.

Informality simply doesn't work well in an environment of large-scale product design and development. More structure is needed to provide the necessary level of consistency and discipline. Each of their stages of design and development are planned and managed with great care. Variations from standardized program procedures may be required, depending on the technologies being included in the new product and the extent of design commonality with products already developed. Since GEAE is committed to staying at the forefront of this business (and no other), the employees become familiar with all of the superstructure and procedures, and this facilitates rather than impedes their product design and development. In summary, GEAE has built and maintains this formal structure simply because it cannot afford not to.

GEAE uses a matrix structure for its new product programs, rather than dedicated teams. This is a common approach in companies that are largely technology-driven because it allows the establishment of technical core groups that are able to stay at the leading edge of knowledge. Such groups simultaneously provide improvements to existing products while they contribute to the design and development of new ones. (Indeed, at GEAE there are always extensive improvement activities around existing engines and such improvements require much of the same process as a new engine.) The matrix structure requires more mechanisms for coordinating work and negotiating agreements than are necessary for a dedicated NPI team where the project (or program) manager has considerable formal authority throughout the product development cycle.

Planning and Managing Projects

Consider the matter of target setting and priorities covered in Chapter 3. A jet engine is composed of a number of components, each requiring great precision, which must come together and form a totally reliable whole. Complexity can be no excuse for even the rarest of failures. Whatever the pressure for faster development times may be, everyone associated with the design and development of a jet engine knows that complete reliability must still be achieved.

Now consider the matter of product safety from a public policy point of view. The U.S. government, through the Federal Aviation Administration (FAA), protects the public from design or manufacturing flaws by mandating that a very rigorous formal test be conducted on the first complete engine that is produced. This testing procedure attempts to simulate actual environmental conditions (although it is clearly impossible to actually fly an engine that has not yet been coupled with an actual airframe): throwing birds into an engine that is running is one of the required steps in this test. The first engine to test (FETT) milestone on the development schedule for a jet engine represents both a source of uncertainty (Will the tests be successful?) and, perhaps, delay (How much time will these tests take out of our target cycle time from bringing the new engine to market?).

For purposes of scheduling, as well as product design, a jet engine is matched to a designated aircraft, which itself is composed of a number of integral parts. As is increasingly typical, time-to-market was critical in the CF6-80A case. From the start, GEAE knew that it would have to reduce a seven-year development cycle time to four. Whereas, the other three cases presented in Part 2 had tight but somewhat negotiable lead times, the case of the jet engine was different; the engine *had to be* available to coincide with the introduction of the aircraft for which it was designed. To do otherwise would be to jeopardize their future competitive viability.

Given these various targets and their priorities, the cost of the engine represents, to some extent, a slack variable. The fact that the engine was designed largely without manufacturing input and then "thrown over the wall" to the manufacturing organization is consistent with this philosophy. Unit costs are managed by a combination of methods. Instead of adopting a mature first-cost philosophy, as did Northern Telecom, GEAE uses a cost projection model that is traditional in many industries—the learning curve. Historical data allow projections of the cost reductions that will naturally take place between the 1st and the 250th engine that is produced for a new model.

GEAE takes an aggressive and purposeful stance on reducing costs. The target cost is aimed at the 250th engine. After the first few engines are produced, the costs are analyzed in detail and an intensive cost reduction effort is launched. In the case of the CF6-80A engine, the cost reduction effort was quite elaborate. Such cost reduction programs have the scale and urgency of an independent product design and development program with monitoring (but not necessarily recertification of the engine) by the FAA.

Developing Collaborative High-Performance Teams

Each new engine program represents a "bet-the-company" situation for GEAE. Their traditional matrix structure makes teamwork a way of life and most of their professional employees are veterans of more than one NPI campaign. Given this context, the notion of high-performance teams is central to the GEAE culture. Increased use of design-for-manufacture techniques and more extensive reliance on simultaneous engineering approaches to cutting lead times will call for even more collaboration across the critical functions of R&D, engineering, and manufacturing.

Coping with Technology Choice and Risk

The GEAE case illustrates many of the concepts presented in Chapter 5 on the subject of technology management. Most notable, perhaps, is their use of a technology pool. To support its targeted work on the design and development of new jet engines, GEAE conducts ongoing R&D on special manufacturing processes, new materials, and product technologies. Thus the various core technologies related to the manufacture of jet engines are continuously being refined, independent of any particular jet engine. Such R&D is based on company efforts to project the competitive trends in the industry. This allows an engine development program to draw new options from the existing technology pool in order to meet the particular requirements and constraints of a new engine program. Technology choice and risk management go well together under this arrangement; program management will deliberately take components that are already proven and combine them with ones that need more design and development to achieve the best overall figures of merit at the least time and cost risk. GE's series of engines within a single product family allow the sharing of a basic design, leading to faster new product introduction at lower costs.

Conclusion

GEAE's design and development effort for the CF6-80A engine was a significant step in an ongoing program of improvement for this industry leader. Despite the complexities of the product itself, the regulatory constraints imposed by the FAA, and dynamic shifts in both the customer and supplier environments, GEAE has managed to introduce successful products in increasingly short-cycle times.

By 1990, GEAE had formalized a program of continuous improvement, and had embarked upon yet another "bet-the-company" new engine introduction, with a pioneering attempt to reduce still further the product development cycle time. This will require even more extensive reliance on simultaneous engineering, use of design-for-manufacture techniques, and earlier freezing of the design (fewer design changes). Despite the continuing pressure for increased speed to market, the achievement of flawless quality must remain the number one priority. Regardless of the complexity of its product and the size of its new product development teams, GEAE is striving to be a nimble leader in its industry.

PART 3

THE BROADER CONTEXT— ENABLERS, LINKAGES, AND CONTINUOUS IMPROVEMENT

So far this book has attempted to provide a comprehensive appreciation for the challenges and opportunities in achieving effective product design and development at the level of the individual project. Many factors and issues were presented in the five conceptual chapters of Part 1 and the case histories of Part 2. The case histories also highlighted the importance of the organizational context within which NPI projects are conducted. In particular, they indicated that new product introduction becomes more effective when a company also develops certain enabling and integrating capabilities, and an orientation toward continuous improvement.

This final part of the book goes beyond the basic frameworks and concepts to explore this broader context of product design and development. The following chapters address these questions:

- What companywide initiatives will enable more effective product design and development?

- What ongoing business activities have primary functional linkages with product design and development?
- How can companies continue to improve their process of new product introduction?
- What issues will they need to address with respect to these enablers and linkages, and in the pursuit of continuous improvement?

Part 3, then presents principles and lessons to guide companies toward an integrated approach to NPI improvement.

CHAPTER 11

INITIATIVES ENABLING
IMPROVED PRODUCT DESIGN
AND DEVELOPMENT

Companies often use corporate staff to develop and promote initiatives in support of their line operations. Part 2 provided several examples of how such initiatives could help strengthen the process of new product design and development: Motorola's Six Sigma program for quality; NeXT's views on the recruitment and development of staff; Northern Telecom and GE Aircraft Engine's use of information technology to communicate design specifications to their suppliers. An integrative view of product design and development needs to include such enablers.

While numerous corporate initiatives might fit within this integrative perspective, these three are particularly important:

- Total quality management (TQM).
- Human resource development.
- Information technology development.

Each of these enablers (see Figure 11–1) is briefly described in this chapter, along with an agenda aimed at promoting more effective product development.

TOTAL QUALITY MANAGEMENT

The 1980s witnessed a true revolution in the world of manufacturing when "Quality" became everyone's watchword. The message that "quality is free" (Crosby, 1979) was liberating. No longer did managers have to think in terms of the extra cost of investing in improved quality. Quality initiatives on the factory floor would more than pay for themselves in terms of the avoidance of unnecessary expenses (and revenue losses) associated with the manufacture of substandard products.

It was not long before this philosophy spread. Under the label *total quality management (TQM)*, companies of all types began to pursue the

FIGURE 11–1
Three Initiatives Enabling NPI Improvement

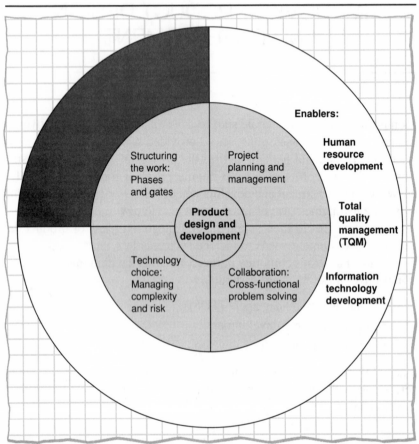

notion that "quality is everyone's job," not just those working on the factory floor. The Malcolm Baldrige National Quality Award, instituted in 1988, is presented annually by the President of the United States to those U.S. companies that have achieved the most exemplary levels of quality throughout their organizations. Thirty percent of the point total in the competition for this award is allocated toward a company's demonstrated performance in being responsive to the customer. It is hard to imagine any company winning this award without becoming proficient in the design and development of products that meet customers' requirements.

The total quality movement shows no signs of faltering. Indeed most large companies probably have a TQM program either in effect or in the development stages. To the extent that TQM is already being emphasized by any company, it will be easier to implement the kind of product design and development that has been described in this book. The reasons for this are both simple and powerful. With respect to NPI activity, a company's prior successful TQM orientation provides:

1. An important signal from senior management.
2. A basis for justifying resource allocation.
3. A value that transcends turf issues.
4. A shared goal for shaping problem solving.
5. Common standards for good work.
6. Tools and methods.
7. The quest for continuous improvement.

An Important Signal from Senior Management

The TQM movement only works if senior management embraces it. By this very act of making the pursuit of quality a top priority throughout the company, senior management is communicating in no uncertain terms: all participants in the new product development process are expected to provide quality service to each other, to others who will get involved with the introduction of the new product and, ultimately, to the customer who will use it. This is an important signal. Without it all kinds of counterproductive behavior (described throughout this book) are more likely to occur.

A Basis for Justifying Resource Allocation

The allocation of resources, human as well as financial, has been described earlier as a key factor in the success of any NPI (new product introduction) project. A company that has already developed a TQM program is much more liable to assess such resource allocation decisions in terms of the probable impact on the quality of the ultimate product and associated services than one that has not. Analysis to justify the allocation of scarce resources to an NPI project will still be required. But the analysis will center around the core concept of meeting customer requirements, rather than narrower traditional financial justifications of expenditures. From a TQM perspective, a new product is more than a discounted stream of revenues. It is also a statement of what the customer can expect from the company in the future.

A Value that Transcends Turf Issues

One of the basic precepts of TQM is doing the right thing, the right way, the first time. This precept is especially important in NPI because it establishes a shared value for all who are involved with this activity. This value encourages team members, the project manager, and those on executive review committees to provide adequate emphasis on the early, formative phases of project initiation and conceptual design. Such behavior is consistent with the economic argument that it costs much less to avoid potential problems of quality than it does to fix the problems once they are already embedded in the product and its system of production. In other words, it is better to spend an extra dollar on prevention, or even $10 on inspection, than to lose $100 (or more) to correct the field failures of a product. The translation of this point of view to the work that is done in the various phases of product design and development is straightforward, and has been described in Part 1 of this book. If all participants in the NPI effort are familiar with these notions from TQM they will be more inclined to adopt this perspective in their own work. This will facilitate collaboration across traditional functional lines, and help avoid unproductive turf battles.

A Shared Goal for Shaping Problem Solving

The TQM movement insists that the company direct all of its energy to the value-added activities that build customer satisfaction. Sustaining the customer's viewpoint throughout the NPI project provides a shared goal that helps structure all problem-solving activities. Using actual or potential customers, marketing information, and field service insights, toward this end, in various ways at different phases of product design and development has already been discussed. Nevertheless, the challenge remains to build such data and information into an integrated set of design and development decisions. A background in TQM helps here because it offers more than an overarching philosophy that quality is everyone's job. It also offers a constant goal, or frame of reference, for shaping issues, options, and trade-offs. Simply put, this facilitates participants working on the same problem at the same time and reaching a consensus.

Common Standards for Good Work

Specific quantitative standards for "good work" have been articulated through TQM programs. Motorola, one of the first winners of the Baldrige Award embraced the Six Sigma standard (described in Chapter 10). Achieving Six Sigma quality is now part of the TQM program at IBM and other companies. Having such a common standard accepted throughout the company can be very instrumental throughout an NPI project as a criterion of acceptance. This is true for all aspects of product design including: the choice of materials; the form, fit, and function of components, subassemblies, and the complete product itself. It also applies to manufacturing and testing processes, including prototypes, pilot runs, and volume facilities, as well as supporting products and services such as product documentation and field service diagnostics.

Tools and Methods

The theory behind TQM, attributed to Dr. W. Edwards Deming, is that undesired variation in productive processes is the source of problems of quality. A set of tools and methods, including statistical process control (SPC), is taught throughout the company as part of a TQM program. These tools are already in widespread use on the factory floor (and in high-volume customer service settings). Use of such information on existing manufacturing and product testing processes can provide valuable inputs in the design phases of NPI projects.

Furthermore, TQM tools and methods could be applied to the product design and development process itself. If this were done carefully over time, a valuable data base would result. Two common barriers to more effective product design and development at this time are the widespread lack of such data and the failure to explore more rigorous application of basic methods of process analysis. This initiative could be linked with an ongoing assessment of NPI projects, as discussed in Chapter 13.

The Quest for Continuous Improvement

Ongoing efforts to improve efficiency and effectiveness are also stressed in TQM. Company programs for TQM emphasize the careful analysis of all critical processes—sets of operational activity aimed at producing a product or service to either an "internal" or "external" customer. Chapter 6 presented a view of ongoing improvement of product design and development through the use of new design technologies. Continuous improvement of the entire NPI process is the subject of Chapter 13.

HUMAN RESOURCE DEVELOPMENT

Over a period of the last few decades, the personnel departments of most large companies have shifted from a preoccupation with rules and procedures for implementing a variety of routine personnel actions to a concentration on "developing the human resources" of the company. Included in this newer charter are the traditional activities of employee selection, training, and career development. Indeed, the notion that "people are our most important product" is receiving more than lip service in many manufacturing companies committed to competing in a world of global competition and continuing technological change.

Ongoing learning by individuals is necessary in the field of product design and development. The need for expertise in the various fields that contribute to the planning, decision-making, and problem-solving activities of the NPI project is both broad and deep. In every specialty—including marketing, industrial design, product engineering, and manufacturing engineering—new approaches and techniques emerge. Existing technical skills need updating. Equally important, the increased use of cross-functional teams demand that participants' interpersonal skills be enhanced. All participants need to learn how to be more effective contributors to the NPI process of their company.

Given this need, and related challenges discussed in prior chapters, the Human Resource Development Organization is faced with four distinct kinds of challenges. Each deals with one of the traditional responsibilities of their field:

1. Setting criteria for employee selection.
2. Developing training programs.
3. Appraising professional employees.
4. Career planning.

Not all of these responsibilities are necessarily lodged in the same group within every company. Regardless of who happens to be in charge of these matters, the challenges, discussed below, remain basically the same.

Setting Criteria for Employee Selection

Improvement of a company's new product development capability begins with its work force, rather than with new technology. The hiring of new employees at all levels therefore needs to reflect today's demands for job enrichment throughout the NPI process. Be it entry level people in design, manufacturing organizations, or their more highly educated counterparts, traditional criteria for employee selection must be modified to include new factors. The ability to work effectively in teams is one of these. Another is an interest in and aptitude for learning about the business and the customer. Basic problem-solving skills, above and beyond any specialized knowledge or aptitude are also important. Functional expertise and preference is, therefore, still necessary but not sufficient for many of the jobs that are key to the successful design and development of new products.

Human resource (HR) professionals charged with setting such criteria need to become very familiar with the current and future challenges of new product development. Only then can they adequately cope with their assignment of translating human resource needs into hiring criteria.

Developing Training Programs

Updated hiring criteria is only one element in a human resource development program. Equally significant are the company's training programs. Three types of training initiatives are clearly needed to enable effective new product design and development: (1) job requirements, (2) interpersonal skills, and (3) team training. A fourth type of training requirement—for strategic skills—is generally unrecognized, but will prove to be increasingly important.

Each of the functional specialists who is likely to play a role in the NPI process should receive some training in the technical aspects of performing well in these assignments. Such topics should be part of the ongoing training provided to members of that business function, be it marketing, design engineering, manufacturing engineering, manufacturing, purchasing, field service, or quality assurance. All such employees need to know how the NPI process generally works in their company and specifically how people in their roles are expected to contribute to the success of that enterprise. Analytic techniques and skills, at the appropriate level of rigor, need to be provided. Exposure to the customer, if not part of the ongoing job activity, ought to be included in such training programs.

Regardless of the substantive responsibilities and training, potential NPI participants need general training in teamwork. Such training typically concentrates on the strengthening of interpersonal skills and group problem solving. Many companies already provide such training as part of their overall TQM program. This dimension of training can be delivered either by individual functional organizations or on a corporate basis, depending on the various managerial agendas at the time. It is not related specifically to the activity of new product development.

Newly formed NPI teams ought to receive a special training program aimed at promoting excellence in their work. Such training would include an in-depth orientation to the company's management control (phase-gate or other) system for NPI, and the roles and responsibilities of all participants. It should also include exercises in building team synergy. Specific training in design techniques and approaches, such as QFD or DFA (quality function deployment and design for assembly, respectively), already in use by the company should be provided to all participants. There might even be a special training module on project management.

Human resource professionals also ought to be promoting the development of strategic skills that build on the others already mentioned. In the field of product design and development, such skills would include:

Scanning the environment to spot emerging trends in technologies and markets.

Identifying needs for alliances in designing and introducing new products.

Setting and achieving project objectives in an environment of change, complexity, and risk.

Empowering teams (including external suppliers) to be more effective.

These types of strategic skills are rare but are clearly needed by executives, project managers, and key specialists. Programs to develop such skills are different from traditional training and will call for innovative approaches.

Clearly, training of the scope and magnitude indicated is an expensive and time-consuming enterprise. Human resource professionals ought to be champions for the development of such training. It is hard to justify such an emphasis on training by trying to answer the question, "Can we afford to do it?" Based on the complexity and significance of new product development for most companies, the proper question to raise and address is, "Can we afford not to?"

Appraising Professional Employees

Establishing employee performance appraisal policy and methods is usually the responsibility of the Human Resource Organization. This is a very dynamic field and most companies try to update their appraisal criteria to reflect current job responsibilities. Since incentives are an important variable in the equation for collaborative NPI (see Chapter 4), it is vital that the right incentives be reflected in the appraisal of NPI participants.

Perhaps the most important criterion to emphasize is the degree to which a participant functions as an effective member of the team. Putting this criterion in place requires that team members receive early orientation and continuing reinforcement of the importance of teamwork to achieving an effective NPI outcome. Formal training can provide an early orientation, described previously. Continuing reinforcement is both a responsibility of the project manager and all members of the team. If a company already has a tradition of emphasizing teamwork, say as part of a TQM program, then its use in the context of product development should be straightforward. However, if NPI is the first business activity where this emphasis on teamwork is being attempted, its successful implementation should not be assumed.

Another issue of employee appraisal arises when dealing with professionals who work on NPI projects on a part-time basis, while retaining ongoing responsibilities in their functional area of expertise. In such situations, the Human Resource Organization needs to develop clear guidelines on how to include performance on an NPI effort as part of the overall appraisal. The role and relative power of the project manager needs to be specified. The trap to avoid is letting the functional manager of the employee minimize the importance of the part-time contributions that were made to the NPI effort, compared to the more visible ongoing contributions to the work of the "home" organization.

Finally, the Human Resource Organization needs to develop an appraisal process that promotes the assessment of NPI participants in terms of overall success in achieving the various team objectives. If the targets of cycle time, development cost, product quality, and unit cost are to be taken seriously, then it ought to be possible to identify which employees were in a position of significantly affecting the achievement of these targets. The appraisal process is thus the critical bridge between the pursuit of NPI targets and rewarding those individuals who were part of a successful collective effort to achieve them. Since everyone likes to be considered to be part of a successful effort and no one likes to be associated with a failure, the project manager, being a close on-the-scene observer, plays a critical role in the implementation of this kind of appraisal system.

Career Planning

New career patterns will have an impact on business processes such as product design and development (see also Chapter 13). The Human Resource Organization can enable positive impacts through their policymaking role with respect to career planning. Two elements of this policy are: (1) the timing and pattern of cross-functional assignments; and (2) the degree of stability and longevity in assignments to those who have skills in new product development. Dealing successfully with each of these elements requires a departure from the traditional pattern of career development in most large companies: those with technical expertise tend to get promoted within their area of functional expertise through several levels of the managerial hierarchy. Now that many companies have experienced radical restructuring, there are fewer levels of management. This should facilitate career planning that places more emphasis on longer term career involvement in product design and development by those who have shown the ability to achieve notable success. Cross-functional assignments, as part of normal career planning, would also strengthen the problem-solving abilities of NPI teams and perhaps allow those teams to be smaller and more productive. Clearly, this is only likely to happen if the employee selection, training, appraisal, and incentive systems are compatible with this approach.

INFORMATION TECHNOLOGY DEVELOPMENT

Corporate information systems groups used to have a relatively simple charter: to oversee the acquisition of centralized computer hardware and to develop or procure batch-processing software applications as needed.

With the advent of on-line data entry, linked networks of distributed computer power, and data base technology, the field of information technology (IT) has broadened and become relevant to almost everyone in an organization. Information technology has become synonymous with communication. Like it or not, we can no longer think only of "people problems" when our needs for better communication become apparent.

IT groups are now positioned to enable much of the technological innovation that supports the process of product design and development. Competitive new product introduction calls for increasing capabilities for accessing, analyzing, and transferring information that has traditionally "belonged" to a limited set of employees. An information-based model of the functional value of various design technologies (see Chapter 6) provides a conceptual bridge for joining IT expertise with NPI applications.

Unfortunately, very few companies to date have gone as far as they might to use IT to help make their NPI process faster, more customer-driven, and better coordinated (across functional, organizational, and geographic boundaries). While isolated applications of IT have enhanced aspects of product design and development, companies would benefit from more extensive support.

One way to think about the payoff from information technology initiatives is the contribution it makes to achieving greater speed to market. The time between the release of a product design and the beginning of volume manufacturing contains a number of information-intensive activities described in Part 1 of this book: design verification and testing, setup of a test process, setup of a production process, and procurement of tools, materials, and equipment. The use of CAD systems with standardized approaches to product design, coupled with engineering data management systems can provide cost-effective improvements in the time from engineering release to production. Through such an application of IT, Allen Bradley's Industrial Computer and Communications Group, for example, reported a recent reduction from 22 weeks to 8 weeks in this interval from design to production of a complex printed circuit board.

Another important potential contribution of IT is the integration through networks of different computing capabilities dispersed among the participants in product design and development. These networks can include mainframes and different kinds of personal computers, disk farms, and electronic vaults (for secure data storage). The technical challenges here include combining multiple architectures, the ability for suppliers to extract data from central files, and ease in reusing existing designs.

Other possible IT contributions could include: more productive development of software for new products through the use of CASE (computer-aided software engineering) tools; the introduction and enhancement of computer-aided manufacturing (CAM) tools; the electronic transfer of manufacturing files (CAD/CAM); and the combination of engineering, manufacturing, and administrative systems (for order entry, billing, and inventory) into computer-integrated manufacturing (CIM) systems.

All of these IT capabilities exist. But which ones should a company pursue? In what combinations? And how fast? IT professionals are often charged with producing a strategic plan that addresses these questions. If this IT initiative is to enable more effective product design and development, the plan ought to be carefully integrated with the operational needs of this set of users. Since the acquisition and implementation of information technology can itself be slow and treacherous, the agenda for action in this area must be developed with care. Issues of technology choice and risk (raised in Chapter 5 with respect to product and manufacturing technology) apply just as well to the adoption of enabling IT technology.

SUMMARY

The enterprise of product design and development is too important and demanding to simply be delegated to those line organizations that have day-to-day responsibility for such work. The work of these organizations needs to be enabled by companywide initiatives to develop a supportive culture (quality) and infrastructure (human resources, information technology).

Over the long term, the enabling power of three corporate staff initiatives can be particularly potent: (1) total quality management (TQM), (2) Human Resource (HR) management, and (3) information technology (IT) development. TQM programs teach participants in product design and development how to embrace shared objectives and set standards for "quality" work. HR programs can strengthen recruitment of people who are well-suited to this field, train them in needed skills, and establish incentives for them to make an extended personal commitment to this field. IT programs make possible the ambitious time-critical, decentralized NPI efforts described in Parts 1 and 2.

CHAPTER 12

KEY LINKAGES WITH OTHER BUSINESS FUNCTIONS

That product design and development is not an isolated business activity should be clear from the conceptual foundations and findings of Part 1 and the case illustrations of Part 2. Its effectiveness directly depends on activities normally performed within allied business functions of the company. The importance of these other business functions has already been mentioned throughout this book and is selectively explored in this chapter.

Here we examine key linkages between NPI and the ongoing functions of manufacturing process development, marketing, and field service (see Figure 12–1). The focus is on information transfer, joint decision making, and supportive actions. With a sharpened sense of such inherent linkages, companies can develop a more integrated approach to introducing new products, one which capitalizes on the full spectrum of knowledge that is already available. Spotting and fixing weak links of this sort are essential responsibilities of senior management. Using these links on a routine basis to increase the likelihood of successful product design and development is the responsibility of all participants.

MANUFACTURING PROCESS DEVELOPMENT

The design and development of manufacturing processes is a complex endeavor, with direct implications for both new and existing products. A full treatment of this subject, clearly beyond the scope of this book, can be found in the recent book of Daniel Shunk (1992). Here, we address the following topics, emphasizing aspects that are tied most directly to the field of product design and development:

1. Process analysis and capacity planning.
2. Process design.
3. Process development.

FIGURE 12–1
Three Business Functions Supporting NPI

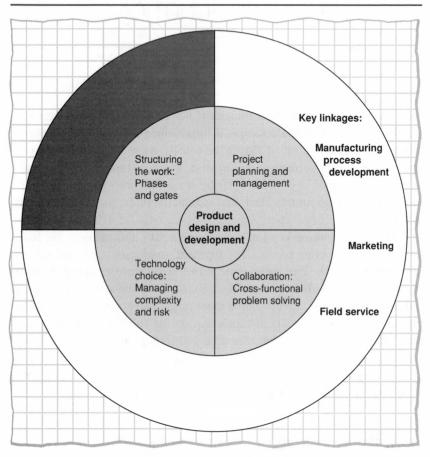

4. Flexible manufacturing.
5. Supplier management.
6. The development/production interface.

Process Analysis and Capacity Planning

In existing manufacturing facilities, industrial and manufacturing engineers and methods analysts routinely examine the manufacturing process. They learn much about the achievable capacity from different kinds of workstations and entire production lines. The knowledge that they accrue is vital to the planning of production processes for new products using similar types of manufacturing capabilities. Even if such individuals are not designated to work on an NPI team, the representative(s) from manufacturing needs to build on this existing base of knowledge when projecting costs, achievable capacities, and work standards.

Process Design

The design of manufacturing processes for new products requires more expansive analysis and problem solving than similar design efforts applied to existing products. For existing products, the specifications are either fixed or are being incrementally modified, a manufacturing process is already in place and serves as a foundation for improvement, a work force already has experience with the product, and the objectives for process design are closely tied to prior experience with that product. In contrast, the design of a manufacturing process for a new product can be a much more open-ended activity with greater strategic implications.

Manufacturing engineers who design production processes for new products make significant contributions at the early stage of conceptual design. Here, as described in Chapters 2 and 6, the need for a close orientation to customer requirements is great. Manufacturing engineers need to become skilled in participating in exercises such as quality function deployment (QFD), where customer-based product specifications are translated into manufacturing issues and options. The ability of manufacturing engineers and product designers to collaborate (see Chapter 4) is a prerequisite to setting product specifications with acceptable manufacturing solutions.

The simultaneous engineering of a product and its manufacturing processes (described earlier) is an especially demanding aspect of new product development. Here, manufacturing engineers must be able to work effectively at process design before the product design has been frozen. Doing this requires a high level of flexibility and a willingness to revisit earlier design decisions as more information becomes available. It also requires skills in anticipating process problems before volume production, or even pilot production, systems have been developed. Constructively interacting with product designers in an ongoing manner around such emerging issues of design is a capability that not all manufacturing engineers will have. The same can be said, of course, the other way, with respect to product designers.

Process Development

The development of manufacturing processes for prototype and pilot testing, although complex, is key to the successful introduction of a new product. Special capabilities are needed to support rapid prototyping, which (see Chapter 2) puts sample products in the hands of potential customers much earlier than traditional modes of product design and development. Learning from such prototypes what the volume manufacturing process ought to look like is another important contribution. Combining an orientation to product performance with issues of manufacturability becomes a central agenda as prototypes are built, tested, and revised. Meanwhile, within many NPI projects, process development is subject to its own critical path, where the design and development of tooling and special equipment become time-critical mini-projects. These are the areas where companies have to decide the extent to draw upon the same people who have process development responsibilities for existing facilities and products, or to assign people only to work on specified NPI projects.

The economics of production, an important part of new product success, depends largely on manufacturing strategy with respect to facilities. Focused factories—dedicated to a limited range of production capabilities rather than making a full line of products—are becoming a more common approach (Skinner, 1974). The decision as to whether an existing focused factory can accommodate a new product should occur as early in the product design process as possible. This early consideration promotes related design approaches, such as the use of group technology (GT), to achieve similarities in the processing needs of different products. Skills in the development of focused factories and the use of group technology reside in the manufacturing organization and need to be made available as needed to NPI projects.

Flexible Manufacturing

Flexible manufacturing is a special variety of production process with special linkages to product design and development. In the 1980s, leading worldwide manufacturers implemented combinations of programmable robots and computer-driven production processes (i.e., numerically controlled machining and automated assembly) to form flexible manufacturing cells that required no hands-on production labor. Some companies went even further by connecting such cells with automated material handling technology to develop complete flexible manufacturing systems (FMS) for producing series of parts or components without human intervention. In the 1990s, the strategy of flexible manufacturing is generally well-understood, and most large-scale manufacturers who produce goods in small batches have had some direct experience with these advanced manufacturing technologies. Two aspects of flexible manufacturing technologies relate to new product development:

1. The use of such *existing* process technology to produce new products in addition to existing ones.
2. The design of new products to be manufactured by *new* flexible manufacturing systems.

The ability to add new products to flexible plants, called *product mix* flexibility was know early on to be one of the potential benefits from adopting computer-based automation (Gerwin, 1982). The development of such flexibility, however, has its own costs and any such flexibility has its natural limitations. From the viewpoint of new product design and development, the important question is: What bounds to product mix flexibility were built into our existing manufacturing systems?

If a new product can be designed to fit within those bounds, the existing facility is a candidate for its manufacture. The advantages in terms of time-to-market and cost of reusing such flexible manufacturing systems are likely to be significant. But the penalties for planning to do this when that manufacturing system is not as flexible as had been assumed can be even more considerable. Companies that already have such manufacturing systems need to test them enough to validate their true range of product mix flexibility. This information needs to be clearly communicated to the NPI team early enough to be considered in matters of product design.

Designing products for new flexible manufacturing processes, quite a different matter, requires careful strategic planning and analysis in the early phases of the project. The manufacturing organization will traditionally be influential for the choices regarding investment in advanced manufacturing processes that are being made. As described in earlier chapters, many companies have a group called Advanced Manufacturing Engineering (AME) with responsibility for such process planning and justification. How this responsibility gets executed within the framework of an NPI project varies from company to company. In many cases an experienced member of the AME group is assigned to the product development team and draws upon the expertise of the AME group as needed. The designated responsibility for achieving the target for unit cost often determines who, in fact, has the final say on new process acquisition.

The main strategic issues in designing products for new flexible manufacturing processes are the types and extent of flexibility that are worth developing, and when this should be done. Resolving such issues is facilitated by the prior formulation of a manufacturing strategy signifying the priority of having high-quality, responsive manufacturing capabilities. To the extent that this is a high-strategic priority, the introduction of a new product offers the best opportunity to commit to the type of flexible manufacturing capability that will serve projected future requirements, above and beyond those of the new product as designed. There is, however, an obvious potential danger in this approach, one which should be considered in Phases 0 and 1 of the NPI project: the risk of missing critical project targets (time, cost, or even quality) when developing and installing such a complex manufacturing capability within the framework of that project. Recall, in the case of NeXT (Chapter 7) that this risk was deemed acceptable, while in the case of Northern Telecom's Norstar project (Chapter 8) a more cautious stance was taken (with respect to the introduction of surface-mount technology).

Supplier Management

Working with suppliers of materials, parts, tools, and equipment—a traditional activity of those who develop manufacturing processes—is a critical part of NPI projects. As a product is being designed, issues of materials and parts arise early. Sometimes the choice of supplier spells the difference between a design solution that is feasible and one which is not. Examples of this sort were contained in the cases of NeXT and Northern Telecom (Chapters 7 and 8). Tooling and equipment are often supplied by external vendors, rather than in-house groups, and require careful specification by those who are designing the overall production process. Integration of the purchasing function into such supplier-related decisions for new products is a special challenge. It calls for Purchasing to play a more strategic role above and beyond the traditional one of monitoring of contract compliance.

Another complex matter is the use of suppliers to manufacture major product components for a new product. In the 1990s, companies of all sizes in all industries are looking more closely at strategic partnerships with suppliers. The four cases in Chapters 7–10 describe different versions of this phenomenon. Although the relative emphasis on external manufacturing is an important element of a company's manufacturing strategy, specific decisions need to be made for each new product being introduced. Making such decisions is not an abstract matter of preference; it requires knowing who the best potential suppliers are and whether a suitable arrangement can be made with one or more of them. The process of reviewing potential suppliers, often called *qualifying the supplier* has become a highly refined activity in this era of total quality management. Moving beyond the point of selecting a supplier, to establishing the optimal role of that supplier with respect to the design of the product, is another important formative decision. Keeping the supplier involved in the design and development project is a challenge for project management. Ensuring that suppliers of material and parts continue to meet conformance specifications and delivery requirements involves Manufacturing and Purchasing.

The Development/Production Interface

Since manufacturing is responsible for the ongoing volume production of a new product, they should participate vigorously in the product development process in order to enable a smooth transition to volume production. Small companies have a natural advantage in this pursuit. They are, by definition, less dependent on strict organizational structures. Fewer people are around, so working relationships among those who do product development and manufacturing ramp-up will probably be closer from the start. The case of NeXT (Chapter 7) is illustrative of what one might hope to find. There, a small group of individuals were responsible for making all decisions of manufacturing process design and they continued to play a significant role throughout the NPI project.

In larger companies these same decisions tend to be made by many different people (Chapter 10, GE Aircraft Engines case). None of these people necessarily will have been closely involved with early stages of the product development team. The organizational solution that seems to be in widespread use at this time is often called the *Initial Production Organization* (IPO). The IPO facilitates the transition from development to production by assigning NPI team members with considerable exposure to prototype development and testing to work directly with the manufacturing personnel who would subsequently take responsibility for the new product. This approach facilitates a smooth transfer of both product and process experience to the manufacturing team. In the future, further integration of roles and responsibilities in the IPO may evolve as product and production experts learn more from each other. Such a movement would make the development/production interface occur even more smoothly.

In summary, the development of manufacturing processes can be a precursor or a response to any new product introduction, or even a well-integrated part of the design of a new product. As a precursor, the analysis of the capacity, quality, cost, and operating ranges and their limitations provide important inputs to upcoming NPI projects. As a response, manufacturing process development enables the market success of the new product. As part of the NPI project, it offers opportunities for setting and meeting more stringent targets (time, cost, and quality). Managers should strive to link manufacturing process development to product design in all of these ways, paying attention to issues and opportunities outlined in this section.

MARKETING

Throughout Part 1, we noted that marketing plays a key role in the NPI process. Those references to marketing were intentionally abbreviated compared with other activities directly impinging on the design and development of a new product. However, in the world of marketing, product design has long been a subject of study and work (Urban, Hauser, and Dholakia, 1987). In this section, we aim to redress some of this balance by elaborating upon connections outlined earlier between traditional marketing activity and NPI.

Anyone who is working on the design and development of a new product would do well to understand and appreciate the roles played by people trained and experienced in the marketing function. It is important to note, however, that in most industries the responsibility for new products is now shared across the operating functions of business: marketing, design (with various specializations), and manufacturing. All key players are being increasingly encouraged to think like customers. While marketing no longer holds sole claim to the customer orientation, it does continue to exert a strong influence on the directions that new products will take. Solid market data, assumptions, concepts, and feedback are critical to the well-being of every new product development effort.

Establishing Customer Requirements

At the conceptual design phase of NPI, marketing has the lead responsibility for identifying customer requirements. Methods for accomplishing this objective vary but usually involve working closely with customers who tend to be among the first to adopt new products. Such "lead users" might, for example, be assembled into focus groups where different product concepts (either simulated or in rough mockup form) can be assessed. The sales force might be polled at this time. Special customer surveys might be designed to create an up-to-date data base on customer requirements. Some companies even contract out for advice from consultants or from special research efforts to probe trends in market needs. Information from all of these sources will be combined and assessed as a major marketing input to the process of product design. Skill by marketing personnel in working constructively with other company experts, for example in areas of industrial design and field service (discussed later), are essential at this point.

Defining Market Objectives and Strategy

Marketing plans for the new product also need to begin at the conceptual design phase. Estimating probable demand patterns, impacts on sales of existing products, and market timing are some of the more traditional analytic marketing efforts at this phase. Other related efforts normally conducted by the marketing group include a competitive analysis that would include a comparative assessment of the characteristics of other related products already on the market. Marketing efforts to position strategically the company's new product concept, based on projections of probable future competitive offerings, are often required at this time. Such analyses are valuable to the product development team because they often offer ideas on how to strengthen product position through the selection of particular product features and functions.

Launching the Product

Marketing and manufacturing groups need to closely coordinate their plans for product launch and production ramp-up. Product advertising and promotion must be carefully timed to create demand when adequate production capacity is available, and not much sooner or later. Early advertising or promotion, or delays in production ramp-up, can generate long waits and dissatisfied customers. Delayed advertising or promotion can be responsible for costly high inventories. Therefore, the cost of the market launch, as well as its timing, needs to be managed since expenses at this point in the new product introduction can be very high, especially for many consumer goods.

When the coordination and cost implications of launching a new product are great, management responsibility is often shifted from the product development team to a product manager (sometimes called a *brand manager*). Other companies delay this transition until some pre-specified period after the initial launch, when the product is well-established in the market. Some companies, such as Northern Telecom (Chapter 8) will assign a senior level product manager at the beginning of the concept development stage who will maintain responsibility throughout the life of the product. Two major advantages are associated with this latter approach: (1) those who took the risks in designing and developing the product get the satisfaction (and rewards) from being in charge when the product becomes a winner in the market; and (2) the dangers of losing momentum and commitment in a transition during the critical launch period are avoided. In any event, it is essential that the product launch be managed by people skilled in implementation as distinguished from planning.

At product launch, marketing is in charge of initiating shipments and then performing market evaluations. They are centrally concerned with whether the customer values the new product and, in particular, with identifying whether product performance matches product requirements. In some companies, the marketing group is even charged with measuring the degree of customer satisfaction with the new product. Specific marketing responsibilities during and after the product launch include: the audit and analysis of sales with respect to market objectives; monitoring customer reactions to the product and its price; and verification of proper product flow through the selected distribution channels. Prompt and accurate data collection is required at this point, coupled with appropriate analysis and recommendations. Marketing also continues to monitor competitive products and sales as part of their ongoing line responsibility. The product launch offers considerable opportunity for rapid learning that could affect adjustments to the product, its mode of manufacture or delivery, or even its price. The marketing organization may be in the best position to suggest that deliberate action is needed in response to unexpected changes in the environment (i.e., customer requirements, competitive responses, technological progress, or the economic climate).

Post-Sales Marketing Involvement

The marketing organization conducts post-sales activity that can be valuable input for product improvement or new product design. These activities include: field trips to early adopters, ongoing focus groups with customers of new products, and analysis of reasons for early product returns. In addition, marketing is often charged with the responsibility of assessing the company's sales capacity to launch new products, and the effectiveness of the distribution, field service, and customer support strategies.

FIELD SERVICE

Field service operations are responsible for the installation of newly delivered capital goods of all types, and then, on an as-needed basis, its ongoing servicing. From a customer's point of view a field service activity is of high quality when it is done smoothly and rapidly with minimum downtime; for commercial customers, the capital goods requiring field service (e.g., electronic equipment) are part of their own business delivery system. Unscheduled losses of the use of the equipment, and even excessive requirements for scheduled preventive maintenance, can be viewed as part of the full cost of having that equipment. For consumer appliances, failures and downtime bring a different form of dissatisfaction.

The determinants of dependable and fast field service operations are the capacity of the field service organization; that is, their ability to dispatch a field engineer as soon as a breakdown occurs, an accurate diagnosis of the cause of the breakdown, and the availability of needed spare parts (or modules). More fundamental, however, to customer satisfaction, is whether the original developers of the product were able to design it to be both reliable and easy to service.

Those who work in the field service organization have a deep first-hand knowledge of the product in use. The field service organization should be involved with the process of new product design and development in three ways:

- Life cycle costing.
- Providing feedback to the NPI team.
- Developing and implementing the service plan.

Life Cycle Costing

Customers care about the full cost of a product from the point of purchase until it is scrapped. More and more companies, accordingly, are building life cycle cost concepts into the process of product design (e.g., see Northern Telecom's Norstar product in Chapter 8). Perhaps, because 85 percent of a product's life cycle costs are determined at the time its functional specifications have been established, service issues must be addressed early in the product development process.

There are many components of a product's life cycle cost that are related to its use in the field; that is, installation, periodic maintenance, repairs, acquisition of spares and associated carrying costs, support equipment, training, and termination (Berg and Loeb, 1990). Accordingly, doing life cycle costing well requires that selected field service personnel be part of the NPI team. In setting targets and assessing trade-offs (Chapter 3), the viewpoints of marketing, engineering, and manufacturing must be balanced by the field service perspective with its emphasis on product reliability and serviceability.

Where might field service concerns tend to be compromised in product design decisions? A dominant trade-off of this type, in this era of global competition, involves the unit cost of a product. In an attempt to save on manufacturing costs, designers strive to reduce the number of parts going into a product. But does the reduction of parts lead to a design where it is more difficult for a field engineer to gain quick access for repairs or maintenance? Another direction for seeking unit cost savings is to substitute less costly materials for some components of the product being designed. But will the less expensive material be as durable or may it lead to breakdowns that eventually add much more to the life cycle cost? Other design decisions aimed at materials savings might lead to more costly repair processes.

Life cycle cost analysis, then, will lead naturally to hearing the voice of field service in the product design process, assuming someone with the requisite experience and insight is included in the cross-functional team. Companies must encourage this kind of active participation from their field service personnel.

Providing Feedback to the NPI Team

Field service representatives spend most of their time on the customers' premises interacting with those who use their product as they perform partial replacements and repairs. These representatives have unique opportunities to know what the day-to-day users want. They are also, in a more limited sense, a customer in that the design of the product affects the ease with which they can install or repair it. Field service, therefore, can be a valuable source of information for enhancing the design of a new product or for improving an existing one. The inputs that they may offer span the phases of the NPI project and can be directed to all of the key team members from marketing, engineering, and manufacturing. The requirements for cross-functional collaboration (discussed in Chapter 4) apply to this aspect of field service involvement.

To the marketing contingent on an NPI team, the field service organization provides critical insights on user needs, the ways in which the product is used, and how customers feel about competitors' products. Such inputs are valuable at the conceptual design phase. They also help marketing to make reasonable estimates of probable warranty costs over the product life, and ensure that market surveys include important questions about field service and customer support. They are in a good position to refine specifications for the serviceability of a new product based on the latest market research.

To the design engineers they provide historical data on reliability problems and their sources. They provide a voice in setting "mean time between failure" (MTBF) performance specifications and other targets related to reliability, availability, and serviceability. They also should be involved in the detailed design phase. Their special interest is in the design of built-in product diagnostics that are becoming common for all kinds of electronic equipment. They can also review product designs in terms of whether they solve the root causes of prior product failures.

Their review would also naturally include an assessment of the accessibility to all parts of the equipment that might require maintenance or repairs. Such reviews may lead to explicit targets for repair times, which can be of vital significance to customers who lose operating capacity whenever such field service is being performed on their equipment. Accordingly, the field service viewpoint argues, based on life cycle costing considerations, to design products to minimize ongoing maintenance requirements.

Manufacturing also needs to draw on field service expertise on issues of safety to the person performing maintenance and repair, operating implications of component selection options, and types of defects expected to emerge under situations of normal and extraordinary use of the product in the field.

Developing and Implementing the Service Plan

The field service organization is responsible for the important "track" of NPI activities generally known as the *service plan*. Based on their experience in high technology industries, Berg and Loeb (1990) recommend that a "Core Team" approach be followed within the field service track of the overall NPI effort. Led by the field service program manager who represents the service function on the full NPI team, this second-level core team would be empowered to develop the full service plan.

The development of a service plan has many elements that parallel those of a new product such as:

1. Defining the service offerings and associated prices.
2. Establishing service delivery standards and technical support procedures (e.g., dispatching operations, customer files).
3. Designing the logistics system for storing and distributing spare parts.
4. Specifying and procuring needed test equipment and possible modifications to information systems.
5. Documentation and training for proper installation and repair of the new product.
6. Providing input for customer instruction manuals.

Implementation of the service plan needs to be carefully coordinated with the schedule for product launch; for example, service personnel have to be trained; service systems and procedures need to be tested; spare parts, test equipment, and service manuals need to be stocked; and field service tracking plans must be put into effect. When the overall product development cycle time is short, the development and implementation of the service plan can fall on the critical path unless it is started early, often before the full set of specifications for the new product have received final approval. The field service organization must also be skilled in dealing across its own traditional boundaries (e.g., service marketing, logistics, dispatch operations, documentation).

SUMMARY

Those who specialize in manufacturing process development, marketing services of all types, and field service operations are an important part of a company's extended capability for competitive new product introduction. Only a few of the people working in these areas may directly participate on the core new product development team, and their participation may be limited in timing and duration. Nevertheless, their contributions are vital because they bring particular expertise and special perspectives to the NPI team.

People from these allied activities need to become familiar with the company's approach to new product design and development, even if their primary work assignment is not in this area. All those who are centrally involved in new product introduction should fully understand the potential contributions of these other fields. Promoting this type of integration in sizable organizations is the responsibility of senior management, who should be prepared to address problems related to the "functional silo" structure of their organizations.

CHAPTER 13

AN AGENDA FOR CONTINUOUS IMPROVEMENT

The paradigm of continuous improvement seems destined to dominate the world of business for years to come. In an environment of dynamic global competition, manufacturing companies that are today's industry leaders do not retain their industry position unless they get increasingly better at all value-added activities and increasingly productive by paring down other "overhead" activities. Those who are not industry leaders must also embrace this philosophy if they are to survive, and they must do so with vigor if they are to have a chance of attaining and sustaining a viable position in their chosen markets.

Most large manufacturing companies have either begun programs of continuous improvement or intend to do so, often under the label of total quality management (TQM), discussed in Chapter 11. Some large manufacturing companies have assigned a senior executive to this initiative and the position of "Vice-President for Continuous Improvement" (or "VP of Quality" or "VP of Customer Satisfaction") has been established in some organizations. Smaller companies, meanwhile, embrace a similar philosophy more naturally, without the self-conscious identification of a program with a special name.

To date notions of continuous improvement have concentrated on the production, distribution, and aftersales service of existing products. This is often a reasonable place to start because customers will perceive such improvements almost as soon as they happen. However, the field of new product introduction deserves similar attention, as does sales, marketing, and business planning. Some forward-looking manufacturers understand that an important objective of current and future new product development projects is to learn how to improve the NPI process itself (see Figure 13–1).

This goal of an effective NPI process will not be achieved easily or quickly in most established companies, where the product is reasonably complex and the requirements are changing. Many basic challenges covered in Part 1—under the headings of structuring the work, planning and

FIGURE 13–1
Working to Improve the NPI Process

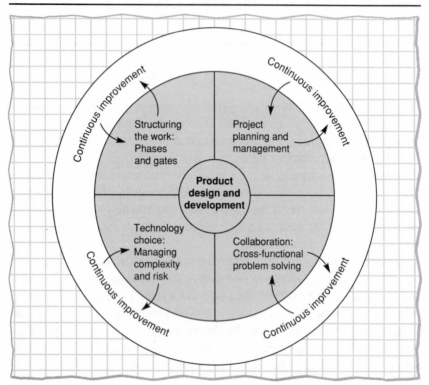

executing the NPI project, creating collaborative, high-performance teams, and managing technological choice and risk—will prove to be more elusive than originally expected. Additional challenges may also arise, as product design and development continues to be the ground for deadly competitive battles. Learning how to be better at this basic business process is something all manufacturers should be doing on an ongoing basis. This includes the many critical informal working relationships —that is, the "white space" not displayed on the formal organization chart.

This chapter addresses two questions with regard to achieving more effective product design and development:

How can management promote organizational learning and continuous improvement of NPI?

What special trends and issues need special attention at this time?

ORGANIZATIONAL LEARNING AND CONTINUOUS IMPROVEMENT OF NPI

Continuous improvement of the product design and development process makes sense, considering its complexity as well as its strategic significance. The many issues of complexity were already discussed. Strategic significance comes from the increasingly short life cycles of existing products. Many companies, driven by rapidly shifting market needs, have a deliberate strategy that in, say, five years, their sales revenues will come in large measure from products that do not currently exist. Other companies, in industries where technologies are moving at a particularly fast pace, look to their new or improved products as the source of rejuvenation when their existing products soon become obsolete. Much of what we generally think of as customer satisfaction in manufacturing has its roots in the design process.

Because the world keeps changing, even those companies that are comparatively successful at new product development are candidates for programs in continuous improvement. Some of the forces for change (mentioned previously) in the NPI process are:

Initiatives by competitors.

Corporate restructuring.

A new market or technology strategy.

Changes in the regulatory environment.

The emergence of new design technologies and practices.

Senior management and certain staff specialists should try to anticipate how such factors might signal the need for changes to the NPI process. Past misjudgments of the effects of such external forces need to be examined as opportunities for organizational learning.

CREATING THE ENVIRONMENT FOR CONTINUOUS IMPROVEMENT

There are natural barriers to applying the philosophy of continuous improvement to the process of product design and development. Some of these barriers also affect improvement efforts throughout a complex organization while others are grounded more specifically in the design and development enterprise. Here, we explore a combination that seems most prevalent at this time:

- Defensive reasoning.
- Turnover of key managers and professionals.
- Design amnesia.
- Weak mechanisms for assessing the NPI process.
- Underutilization of structured design technologies.
- Breaks in the new product "rhythm."

Tempering the Tendency of Defensive Reasoning

Recent experience with continuous improvement suggests that while managers and professionals are becoming enthusiastic about continuous improvement, they are often the main barriers to its success. The reason, according to Chris Argyris (1991), a noted scholar of organizational behavior, is that defensive reasoning at the individual level is a hidden barrier to continuous learning. Successful managers and professionals are handicapped by their lack of experience with failure and how to learn from it. Their fear of failure (and associated loss of job level, power, or even employment) promotes defensive reasoning, in which other people and external situations are blamed for failure. Productive reasoning, rather than defensive reasoning, is a key prerequisite for continuous improvement.

In the context of product design and development this means not blaming the ruthless competition or the fickle market for the failure of a new product. Instead, the productive question is what the NPI team could have done better and when, in the process, this should have occurred. A similar stance needs to be taken on matters of internal management and organizational participation. Benchmarking the NPI process and setting objectives that will stretch current performance can be effective stimuli to prudent risk-taking and learning.

Senior management needs to help employees change their familiar reasoning patterns so that they can be more open to the data-based analysis of the NPI process and its outcomes. Only then can NPI project managers and team members constructively identify the lessons that will lead to improvements in the design and development process. Since senior management may themselves be victimized by patterns of defensive reasoning, it becomes an especially serious challenge to initiate a process through which key employees can learn to learn.

Dealing with the Turnover of Professional Employees

The rapid movement of professional people in corporate America can be a serious barrier to continuous improvement. People who perform well and accumulate valuable skills and experience either get promoted to higher levels of management as a reward for their accomplishments, or they move on to take positions of greater responsibility elsewhere. Even if they stay in the same line of work, they are liable to be reassigned to another geographic region or product line. Such patterns of movement by professional people mean that it can be hard to learn from past failures and even to replicate past successes when companies lose the skills and expertise that they have developed. In the short term, however, efforts to promote continuous improvement in NPI will have to accept such turnover as a reality. Therefore, new approaches, designed to handle the likely turnover, must be created.

As a form of captured learning, carefully designed training programs offer a partial antidote to the plague of professional turnover. Chapter 11 discussed the types of training that would support any team-based approach to product design and development: depth of skill, cross-functional sensitivity, the structure of project management (phase and gate systems), and team problem solving. Additional types of training might be required to overcome deficiencies from normal turnover of staff and/or management. For example, a successful project (or program) manager might be assigned to develop and lead a short course on "effective project management" including such topics as:

Participating in team selection.

Guiding a team through tough spots in the design/development sequence.

Anticipating issues that can arise in executive gate reviews.

Negotiating with those external to the project team on issues of priorities and resource allocation.

New members of the Executive Review Committee might be required to attend a short program in which past and current members review the role of the committee and the types of questions, reactions, and decisions that have proven to be most effective in guiding (but not interfering unduly with) the work of product design teams. The most effective courses of this sort are those that include the generation of an action plan for implementing improvements.

The selective use of informal project advisers is another device for coping with excessive staff turnover. Depending on what type of experience is lacking, senior management could designate one or more advisers to the project. The criteria for selection of the adviser(s) should include interpersonal characteristics as well as desired substantive experience and skill, because serving in this kind of informal role can be a delicate matter. The role of the adviser would have to be specified with care to avoid confusion over the nature of the contributions that would be expected and the priority that this part-time and temporary assignment would have compared to that person's ongoing job responsibilities. A possible side benefit gained from selecting a well-respected person as the informal advisor is to boost the morale of team members by generating the attitude that the project is indeed an important one.

Protecting against Design Amnesia

Producing one winning product design is not enough. In the spirit of continuous learning, a company must be able to capitalize on design breakthroughs that occurred in prior product introductions. Here we are referring to the process of achieving a novel design, rather than any particular design feature or function. The inability to do this is a corporate malady that has been termed *design amnesia* (Ealey and Soderberg, 1990).

This malady is clearly connected to the symptom of staff turnover previously discussed. If nobody who participated in the last design breakthrough is assigned to the current project, it is likely that design amnesia will arise. But the opposite does not hold: even with some of the same people involved, new product designs may not contain adequate memory of past lessons.

One way to protect against a virulent form of design amnesia is to encourage team members to draw heavily on the accumulated expertise in their respective areas of specialization. In each specialty of design (e.g., industrial, mechanical, electronics, software, or automation) a person or small working group would be responsible for maintaining records of the collective expertise that the company or a key supplier has accumulated in that field of design. A test of the validity of this approach is whether such persons are able to discover which of the "new" design ideas have already proven tractable or intractable in prior and comparable design situations.

Another effective way to protect against serious design amnesia is to assign members of current design teams to subsequent new product development projects in that product line. These individuals should be encouraged to bring lessons from their past design efforts to their new assignments. If this process of verbal transfer is well-managed, past experiences can naturally flow into the new project's design considerations. To make this strategy viable, management must cope with the two challenges of defensive reasoning and turnover of professionals outlined above.

Adopting Structured Design Technologies

Continuous improvement in NPI involves more than drawing on people with experience and inventing other devices to retain lessons from prior successful practices. It also depends on the ability to adopt new technology and practices that promote a more structured and consistent approach to product and process design. Chapter 6 presented a model of how a wide range of design technologies and practices can enhance various functions basic to any NPI effort. The agenda for continuous improvement here is capitalizing on the potential of such technologies and practices to achieve a more effective process of product design and development.

The dual approach is to get better at using the existing technologies and to selectively adopt new ones as they become available. A relatively routine aspect of learning deals with the technology itself. What are its features and functions? How do they fit with other tools, techniques, software, or data bases that people are already proficient in using? These kinds of questions can be answered through special assignments to employees already familiar with the way that product design and development is being done by the company. Such assignments should be viewed as learning exercises that examine the way that various NPI activities are actually performed, as distinguished from the formal (or textbook) version.

A more fundamental learning challenge is to identify how a company can adopt and use new design technologies more effectively. In general, the transfer of any process technology from one user group to another is unlikely to be routine. Design technologies—such as quality function deployment (QFD), computer-aided design (CAD), computer-aided manufacturing (CAM), and design for assembly (DFA)—can be expected to raise significant implementation challenges. Learning how to implement design technologies more smoothly, with less disruption to the productive work of the organization, and to use them more effectively, with greater positive impact on the design and development of products, should be an ongoing part of a program of continuous improvement. Anticipating the operational implications of new applications software packages that are procured from external suppliers can be especially challenging (Rosenthal and Salzman, 1990).

Routinely Assessing Completed NPI Projects

Organizational learning builds upon individual learning if the proper mechanisms are established: an ongoing capacity for recording what individuals have learned coupled with formal exercises in comparing and contrasting such learnings and making them available to others. A start in this direction is to require project team members to keep notes (perhaps in diary form) of key events and decisions, as they perceived them, throughout the life of the project. Such descriptive note-taking is quite different from the traditional progress reports tied to project schedules. To be valuable, this note-taking exercise should be done in the spirit of learning, rather than to promote subsequent defensive reasoning.

With such notes and other project documentation as background, routine assessments of all completed NPI projects can be conducted to promote needed organizational learning. Such project assessments, often called *postmortems*, should routinely receive the attention of senior management, regardless of whether the new product was a market success or a failure. The purpose of the exercise is to review the design and development process in light of its near-term outcomes.

The assessments should look deeply into the sources of both effective and ineffective outcomes, from the point of view of internal company action or inaction. The ultimate question to be asked, based on such analysis, is: What should we do differently in our next new product development project? Potential answers to this question could aim at any of the areas discussed in Chapters 2 through 6. For example:

How to impart a customer-driven attitude that will naturally improve products, the NPI process, and subsequent service.

How the formal phase-gate review process ought to function.

What the role of the project manager ought to be.

How to set performance targets for the NPI effort.

How to manage the project to achieve these targets or, when this appears infeasible, to modify them.

How to provide a workplace culture that is more supportive of cross-functional collaboration on NPI projects.

How to anticipate and then manage the element of technological risk as it applies to the design of a new product.

How to use new design technologies more effectively.

The analysis over time of completed projects will shed light on all such factors, although for any single project the lessons may be quite restricted in scope. The important thing about such project reviews is that all key participants genuinely try to confront the issue of how the NPI enterprise could be improved.

There is, of course, a time and a place for celebrating the completion of a successful project. But this should not be allowed to obstruct the process of continuous learning. Nor should the feat of admitting failure stand in the way of the learning process. Human nature and organizational politics, as already mentioned, often pose obstacles to the self-evaluating organization, but recent progress in the movement toward organizational learning provides grounds for optimism when this is attempted in a constructive manner.

Developing a New Product "Rhythm"

Some companies with long-term records of success in product innovation develop a rhythm characterized by streams of products appearing on the market at a high and consistent frequency. This rhythm is detected in retrospect, when looking at the pattern of new product introductions and the timing and relationships between successive new products. Such a rhythm does not arise randomly. Rather, it is the result of a deliberate strategy and intensive work aimed at predictably propelling a company from one NPI to another.

Part of the notion of rhythm is the regularity with which products are introduced to the market. Increasingly short product life cycles in many industries require that companies be well on the way toward releasing the next new product just when their current product is being ramped-up for volume production. This means that deliberate overlap needs to be planned into successive product development cycles. This phenomenon is shown schematically in Figure 13–2.

Achieving an NPI rhythm is especially difficult due to the dynamic character of markets, competition, and technology. Market needs develop at a somewhat unpredictable pace and direction. Likewise, competitors deliberately try to introduce new surprises, either in terms of the new products they offer or the timing of their release. Finally, unexpected progress or delays in the evolution of key areas of product technology or manufacturing processes will occur. Companies of all types will have to make a deliberate effort to learn more about the competitors' NPI, manufacturing, and service capabilities. They will also have to improve forecasting capabilities and use environmental factors and trends in their continuous improvement initiative for product design and development.

Achieving an NPI rhythm cannot simply be mandated by senior management for prompt adoption, but must be considered as a goal to be

FIGURE 13–2
NPI Rhythm: Regular Intervals for Introduction

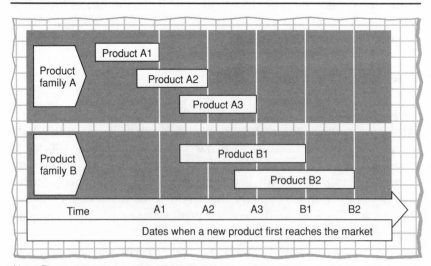

Note: Bars represent complete NPI cycle time for each product.

pursued over an extended period of time. Achieving this goal calls for coherent product planning, close links between R&D and product development, and a discipline in resource allocation, as well as all of the capabilities for effective product design and development that are discussed in Part I. A related requirement is the review (sometimes called Gate 4) of existing products and removal of obsolete ones from ongoing production, thereby pruning the portfolio of products being offered at any time. Even if a company's market, competition, and underlying technologies remained fairly stable, developing and sustaining an NPI rhythm would require continuous improvement as impediments were identified and remedied.

SPECIAL TRENDS

An agenda for the continuous improvement process should include those special trends that are likely to influence effective product design and development in years to come. Part 1 occasionally made passing reference to emerging trends but intentionally avoided lengthy sidetracks. Here we deal with seven such subjects that have emerged as important ingredients of long-term success in new product development. As they become more significant and more pervasive, these trends are good candidates for widespread experimentation and organizational learning.

Three of these trends have to do with the process of managing new product design and development. The other four deal with the scope and contents of design decision. They are:

1. NPI Management Process:
   a. Continued pressure to reduce time-to-market.
   b. Increasingly global business orientations.
   c. Improved performance measurement.
2. Product Design/Development Decisions:
   a. The changing manufacturing work force.
   b. The strategic use of industrial design.
   c. Legal issues in product design.
   d. Environmental issues in product design.

These subjects are already on the agenda of leading companies in different industries. One can predict with some degree of confidence that in the next few years many more companies will be grappling with them. Each of these subjects will require attention from a wide spectrum of managers within any single company. Each has the potential to stimulate radical change in the way that product design and development occurs. Companies that make the most progress on them will be well-rewarded. We now provide some background on each of these subjects, identify the critical management issues, and suggest fruitful directions for improvement.

Continued Pressure to Reduce Time-to-Market

There is no reason to believe that the pressure for speed-to-market—coupled with increased quality and customer satisfaction and reduced product cost—will subside in the foreseeable future. As one NPI team completes the commercialization of a new product in record time, customers will demand that the next product be designed and developed even faster. This will be true regardless of the relative complexity of the product or the length of its development cycle. In Chapter 10 we described a case where GE Aircraft Engines was able to reduce a seven-year cycle to one of four years and they now have set a goal of reducing the "front end" of that cycle (from initiation to first-engine-to-test) by 50 percent. Simpler products that used to take several years to develop have been reduced to periods of about 18 months and many are already aiming at well under 1 year. At the same time that improved performance of the NPI process is being demanded, external expectations by the customer are also rising. Given such trends, what can we expect from such unrelenting pressure to reduce the product development cycle time?

Some companies will learn how to be more productive in their use of time and, as a result, will become more competitive. They will plan on achieving a reasonable amount of simultaneous engineering and eliminate wasted time that comes from poor scheduling or excessive management review. They will become skilled in planning the introduction of product families, thereby embodying less technological change and risk in any single new product introduction. They will become more skilled in producing manufacturable designs, in rapid prototyping, and in the smooth transition from development to manufacturing. They will test market and develop product distribution channels that will be tightly scheduled with product design and development activities. In short, they will make a comprehensive and purposeful effort to place a shorter cycle time into the overall product introduction equation. By learning how to integrate the customer throughout the NPI process, they will also manage to retain the discipline and focus necessary for competitive success.

Other companies, in contrast, will impose arbitrary and overly ambitious cycle time targets without learning how to be more effective in product design and development. Many of them will fail in the marketplace due to some combination of delivering new products later than announced, rushing them to market despite serious deficiencies, or failing from the outset to understand emerging customer expectations. The conceptual perspectives and findings presented in this book are intended to help prevent such undesirable outcomes.

Increasingly Global Business Orientations

The continuing trend toward global companies and markets, discussed by Ohmae (1990) and others, has implications for the management and conduct of product design and development. Some of the implications of doing new product development by involving participants from different locations, organizations, and cultures, were already introduced in Chapter 9, in the case of Motorola's Keynote pager.

By definition, global NPI efforts cannot have fully co-located project teams. Following a highly disciplined NPI process promotes more communication and joint problem solving by the dispersed members of a design and development team than would otherwise occur. Learning how to rely on that process, supplemented by advanced communication technology and prescheduled, face-to-face interaction, will require considerable experimentation and assessment. However, there is the potential to achieve shorter project lead times if design and development activities can be structured to utilize the 24-hour day available to globally dispersed teams.

Managing across organizational boundaries becomes especially challenging in a global context. This is as true for widely dispersed plants or engineering groups in a single company as for external suppliers from another country. Interorganizational partnerships, both internal and external to the company, need to be crafted with special care when it is clear from the outset that day-to-day interaction will be limited by geographical distance. Agreeing in advance on the purpose and desirable extent of this interaction is a good place to start. The selection of partners who have a successful track record of performance under conditions of global dispersion will reduce the risk inherent in this approach. Structuring the partnership in a way that maximizes the focus that goes into the project and then minimizes the required interaction is another approach, although this may not be feasible for some projects.

The potential problems of different workplace cultures, as identified in Chapter 4, can be severe when participation in the design and development is truly global. It is more helpful then to think about effectiveness under conditions of cultural differences, than to strive to develop a uniform culture. The first step in so doing is to appreciate as fully as possible the differences that do exist. Some will be obvious, such as language and leadership styles. Others will be more subtle, such as willingness to take risks or attitudes toward deadlines. Japanese NPI teams, for example, seem more comfortable doing extensive product and project planning than is typical among their U.S. counterparts. Project and program managers need to take account of those differences both in planning and executing the various phases of product design and development. Self-directed work teams can help to bridge cultural and behavioral differences and may use these differences as strengths.

The trend toward more global businesses also has significance for new *products*, as distinguished from the *process* of designing and developing them. Products aimed at global markets raise special design issues. Design teams need to become adept in dealing with new trade-offs in matters of labeling, packaging, and documentation. They need to apply principles of modular design to handle needed variety in power units and to meet varying technical standards. They need to be more universal in their consideration of ergonomics, styles and conditions of use, product safety, and aesthetics. In other words, the practice of design and, perhaps, the composition of the design team, need to adapt to this new reality. Few companies, if any, have mastered this challenge, although many are beginning to face it.

Improved Performance Measurement

The 1980s brought a growing realization that the world of manufacturing was suffering from a counterproductive system of performance measurement. When standard accounting practices are used as the primary approach to managerial control, long-term, cross-functional perspectives are difficult to sustain. Observers came to the conclusion that it was time for companies to "cut the Gordian knot" that bound them and to develop performance measures that are better aligned with the companies' strategy and actions (Dixon, Nanni, and Vollmann, 1990). By the early 1990s, it was clear that many leading companies were struggling to establish improved measures for managerial use, ones directed more to external rather than internal aspects of performance.

For those engaged in the design and development of new products, this emerging capability will provide a basis for strengthening the collaboration in teams. Measures aimed at the effectiveness of the NPI process and its ultimate outcomes can be used to establish incentives for team members to share the common overall objectives of the project. Over time, these new measures will change the perspectives and behavior of team members who currently tend to embrace their own traditional, and more limited, functional orientations in defining and solving problems of design.

The theory of performance measurement points to the directions for change, but once new measurement systems are introduced, they, too, need to be assessed and improved. Part of the learning opportunity inherent in any single NPI project is the extent to which the performance measures seemed to provide favorable guidance. Two types of shortcomings need to be identified: (1) clarity of the metric, and (2) application of the metric. Clarity of the metric deals with the degree to which a relevant concept is made operational. For example:

> If development cycle time is deemed to be one of the important performance measures for an NPI project, has the appropriate starting and ending point for "cycle time" been defined?
>
> Is this a metric that will be clear to all team members and managers associated with such projects?

Application of the metric deals with the collection and use of data.

> Can consistent methods be applied to collect accurate data on the desired performance measures?
>
> Can these measures be compared across projects to indicate improvement of the NPI process?
>
> Are these measures useful as a basis for rewarding outstanding performance?

The continuous improvement of new performance measurement is apt to be the responsibility of someone skilled in the design and imple-

, Time to Volume mfg = key
1st working si.

mentation of accounting systems. This effort will also require the involvement of NPI participants and managers, since they can be the source of insights on the clarity and application of existing metrics. Ultimately, it is most critical that the metrics be customer driven.

Once a new performance measurement system has been put in place, it becomes an important ingredient in the continuous improvement of the company's NPI process. Data from this measurement system should be used as part of a routine, in-depth assessment of each project (as outlined above). Here one should compare actual performance with targets set at the beginning of the project. The targets should include factors covered in Chapter 3 such as development cycle time (and meeting key milestones); development cost; unit cost (initial and experience curve where appropriate); conformance to manufacturing specifications (yield, ECOs); conformance to design specifications (functionality, size, weight); reliability; service requirements; warranty cost; and—perhaps most important— overall customer satisfaction. Some of these measures will be available at the completion of the project, when the product has first become commercially available, while others will have to wait for a reasonable amount of time in use. Precise measures and when to take them will vary from industry to industry.

An ongoing comparative analysis of key performance metrics over time should be conducted independent of any single project assessment. It does not deal with specific aspects of a product or its manufacturing process. Instead this analysis looks at aggregate indicators of time, cost, and quality. Questions to be asked in such analytic exercises are: Are we getting better? How do we compare with our major competitors? Where should we be placing the most emphasis on seeking improvements? Such comparative performance analysis should include some metrics that are oriented to product families and product generations, not just individual projects.

The Changing Manufacturing Work Force

Much attention has been placed in recent years on the changing manufacturing work force. Projections for the United States indicate that new workers entering the U.S. labor force have a different demographic distribution from the work force that existed in 1985 (see Figure 13–3). An increase in the number of immigrants will add to communication difficulties in both manufacturing and service job settings. Furthermore, there

FIGURE 13–3
The Changing U.S. Work Force

Work Force	Labor Force, 1985	New Workers, 1985–2000
Native white men	47%	15%
Native white women	36	42
Native nonwhite men	5	7
Native nonwhite women	5	13
Immigrant men	4	13
Immigrant women	3	9
	100%	100%*
Total work force	115.5 million	25 million

*Does not equal 100% due to rounding.

Source: Hudson Institute (Bennett, 1989).

is no reason to believe that the literacy rates in the United States are likely to improve, and it is already much lower than, for example, Japan (Nussbaum, 1988).

While these changes will not occur overnight, those responsible for designing new products should begin to take this work force factor into account. In particular, they must learn to avoid product designs that unintentionally place unachievable demands on the work force. This type of thinking, fortunately, has already begun under the label of "design for manufacturability"; the technique of design for automated assembly (DFA), as described in the appendix to Chapter 6, simplifies the assembly process so that it could also be performed more efficiently and reliably through manual methods.

Formal techniques for assessing proposed designs for their manufacturability are not a panacea because the definition of this term is itself dynamic. As the projected changes in the work force begin to materialize, even the definition of what is manufacturable with touch ("hands-on") labor will change. This determination will be made more difficult given other concurrent changes likely to occur; that is, the use of newly developed materials, requirements for tighter tolerances on existing manufacturing processes, more advanced equipment requiring different and higher level skills.

Another probable implication of the changing work force is that management will demand the more extensive use of flexible automation. After all, one of the justifications for designing a new product for automated production processes has always been to achieve levels of consistency and accuracy that would otherwise be difficult. But the lessons of the 1980s will not be soon forgotten: flexible manufacturing requires flexible people (Graham and Rosenthal, 1986). Totally automated flexible manufacturing systems are now the exception rather than the obvious wave of the future. To some extent, most manufacturing processes will depend to some extent on human skills. Those who design new products and their associated production processes will need to avoid making naive assumptions about automation and be aware of the skill base of the work force.

These complexities all support the argument that production workers and their supervisors must be more directly involved at critical points in the NPI process. To begin, they might be given an opportunity to react to early plans for the upcoming manufacturing process. In addition, more emphasis might be placed on preparing realistic plans for achieving needed work force preparation through formal training and participation in pilot manufacturing. To the extent that these activities are not part of the current NPI process, they should be introduced as part of a process of continuous improvement.

The Strategic Use of Industrial Design

Industrial design was described in Chapter 4 as a visually oriented discipline that can add great value to the NPI process. Depending on the critical success factors for a new product, industrial design can be focused toward different ends. For example:

> Using a new, less expensive material in an attractive and functional manner, when the new product is aimed at a lower price than the company's prior products.
>
> Using design to create a special emotional feel for a product, to fit a unique market niche (e.g., the Mazda Miata).

Industrial design can also be used more strategically, in ways that their potential contributions have been relatively untapped. Examples of good design having such indirect impacts are:

1. Corporate image and employee morale are raised by winning design awards, even if they are for future product concepts.
2. Outstanding communication design (i.e., packaging, instruction manuals, advertising, and so on) adds perceptual value to the tangible product itself.
3. The design of environments (such as offices, showrooms, and retail outlets) can enhance the effectiveness of work performed in those settings.
4. A coordinated program of design for corporate identity (e.g., signs, logos and other artifacts) affects public impressions of the company as a whole.

In these and other indirect ways, industrial design resources can be used to achieve competitive advantage, above and beyond their direct contributions to the success of any single product. Companies need to learn how to use design resources in strategic ways that are responsive to market conditions and consistent with corporate strategies.

Legal Issues in Product Design

Seeking patent protection has been a traditional role that lawyers have played in many kinds of new product introductions. Until the 1990s, however, this protection has been aimed almost exclusively at utility patents that refer to the operational and functional characteristics of a product. Design protection may also be sought through the application for design patents that refer to the appearance of a product.

The key issue in design patents is whether the product is distinctive and the design nonfunctional (i.e., a shape or surface decoration). One of the criteria for such patent protection is that there are other designs that would achieve the same product functionality. Accordingly, industrial designers may be called upon to visualize some of these other alternatives and create models for review as a basis for securing the design patent. Companies seeking design protection must be careful in their advertising not to make any functional claims for the product that are determined by its design. This relatively new area of patents offers considerable opportunity for organizational learning.

Another legal dimension of product design is the products liability law, which deals with issues of safety and reliability. The difference between reasonable use and misuse of a product is central to questions of liability. This places a premium on designing a product with explicit attention to conditions and styles of use. Ergonomics (Chapter 4) can make important contributions here. Juries in liability cases can decide, even when a product was clearly misused leading to an accident, that the misuse should have been anticipated and the danger removed through deliberate design solutions. The same can be said of liability claims based on product failure through more demanding conditions of use than considered by the product designers. Companies must learn how to avoid potential liabilities through more aggressive and anticipatory design activity.

Environmental Issues in Product Design

The industrialized world is showing signs of serious concern about protecting the environment, and this concern is beginning to be felt by manufacturers at the stage of product design. Undesirable environmental impacts can arise from design decisions regarding product features, shapes, materials, assembly, processing, and packaging.

Responding to regulatory mandates, the automobile industry designed the exhaust systems of cars to reduce air pollution. More recently, this industry has employed advanced CAD/CAE (computer-aided design/computer-aided engineering) techniques to design more aerodynamic shapes for lower fuel consumption.

Industries of all types are becoming increasingly conscious of the environmental impacts of production processes—such as PCB (printed circuit board) cleaning—through which new products are made. As pressure mounts to reduce the type and extent of pollution to the air and water arising from existing processes, products and processes will be redesigned with this objective in mind. In addition, the potential environmental disruption caused by EMI (electromagnetic interference) and RFI (radio frequency interference) in electronic equipment can be reduced or eliminated through simulation of PCB designs and the testing of prototype boards.

"Green packaging" initiatives are also on the rise. The challenge here is to design a package that is less environmentally damaging, while also achieving marketing advantage through quality design. One type of initiative is simply to reduce package volumes for a variety of consumer products. A more fundamental attack on the problem is to design recyclable packages, enabling companies to save on materials while reducing the requirements for solid waste disposal. Kodak's single-use camera with factory-installed film was redesigned—and subsequently renamed from "Fling" to "Funsaver," which is a more environmentally appealing name—so that it could be opened, recycled, and then reassembled with film by Kodak for subsequent resale. In this case, the film processor serves as the recycling agent.

Another opportunity for environmental responsibility in product design is the emerging practice of design for disassembly. As the name suggests, the goal of this procedure is to design products in ways that facilitate their ultimate recycling, after their useful life is complete. This criterion may conflict with some of the other structured design methods as it leads to the physical characteristics of parts and how they are joined. Companies in durable goods industries where recycling holds great rewards in terms of environmental protection would do well to begin experimenting with this new approach and then to include it as part of their program of continuous improvement.

SUMMARY

Improving the design and development of new products has moved upward on the agenda of most manufacturers. Change efforts are widespread, and have gone on long enough for many companies to see real differences in their process and their results. For many companies this process of change began with major restructuring of the NPI process. Gate review systems were instituted and the phases of work were specified more formally with newly specified deliverables and related decision criteria (Chapter 2). Project planning became more rigorous and the role of the NPI project (or program) manager got redefined (Chapter 3), usually in the direction of greater influence over team selection and leadership. A new emphasis was placed on the use of cross-functional teams (Chapter 4), reducing technological risk (Chapter 5), and steps were taken toward the use of more structured approaches to product design (Chapter 6).

These companies have already begun to learn how to be more effective in creating new products, and their managers have embraced this process and their new experiences with great enthusiasm. For such companies, the stage is set for a more long-term view of learning and improvement with respect to product design and development. Companies that have not yet launched such changes but plan to do so would be well-served to adopt the learning perspective from the outset.

Continuous improvement for more effective product design and development is as important to the well-being of a manufacturing company as is any single new product success. Companies that aspire to prominence in their industry have no option but to take seriously the challenge of organizational learning. Such learning needs to be a top priority or it simply will not happen. It requires a commitment to overcoming problems of defensive reasoning, staff turnover, and other sources of design amnesia. It usually requires more aggressiveness in adopting new design technologies and practices that facilitate organizational learning. It also requires building upon the project orientation that we have emphasized and seeking a productive rhythm in the introduction of new products over time.

Three aspects of managing the process of product design and development seem particularly appropriate for inclusion in a program of continuous improvement. Companies must learn to meet continued pressure to reduce time-to-market without sacrificing other critical project objectives. Many must learn to take advantage of the opportunities and deal with the complexities of product design and development in a global context. Finally, companies need to learn to measure performance in ways that promote effective cross-functional, team-based NPI projects.

Four aspects of designing products will need ongoing emphasis and improvement. Two of them deal with the effective use of key resources: industrial designers and the manufacturing work force. Two others deal with emerging realities of the marketplace: increasingly complex legal ramifications of product design; and growing concerns for the damaging environmental impacts of product designs. An awareness of these resource and market-related aspects of design, and an ability to consider them at the right time in the NPI process, will be critical ingredients of success for many companies.

CHAPTER 14

CUTTING LEAD TIME AND INCREASING CUSTOMER SATISFACTION

Successful companies view effective product design and development as a journey, rather than a destination. They expect competitive pressures to drive standards of performance ever higher. While this journey needs to be taken by many people in the company, senior management is responsible for charting its direction. These managers should continue to look for visions of behavioral, organizational, and technological improvements to this process. They should also view current new product introductions as opportunities to build strategic design and development capabilities for the future.

Stating that new product development is a priority is a crucial step toward achieving it in established companies, because changes of this magnitude are hard to accomplish without organizational commitment. When competitive pressure is undeniable, fear is easily mobilized to drive change. However, when competitive pressure is not yet apparent, management needs to provide a long-range vision and strategy, coupled with supporting incentives. People who know what they are supposed to accomplish, why they are supposed to accomplish it, and how they are to proceed will change their behavior, given reasonable incentives to do so.

Our examination of the NPI process has incorporated different perspectives and raised many issues. It has also provided insights on the kinds of changes that are generally needed, and why such changes are likely to result in shorter lead times (or cycle times) and more satisfied customers. The summary that follows outlines the first steps in the journey of effective product design and development. It emphasizes how behaviors need to change between senior management and others in the organization (vertical relationships) and also among those at similar levels across functions (horizontal relationships).

SIMPLIFY VERTICAL RELATIONSHIPS AND EMPOWER THE TEAM

Excessive vertical interactions in the process of product design and development can cause delays and even contribute to product failure. Senior management, whose responsibility it is to determine the nature of these vertical relationships, should try to avoid such undesirable outcomes by structuring the work for early idea validation and business justification, becoming committed to an NPI project, and then concentrating its involvement on the role of facilitator.

Structuring the work requires that adequate resources be allocated to the early phases of a project, when customer needs are identified, project targets are set, design concepts are explored, and issues of technical feasibility are addressed. Providing more attention to these early phases tends to reduce or even eliminate the need for subsequent changes to the product specification, thereby diminishing the overall development time and cost.

Timely management attention is most critical at the early points of review leading to the validation of a new product idea, and the preparation of a business proposal and a project plan. At the successful completion of these initial reviews, management should be confident that the new product will be compatible with market requirements and company strategy and that its plan for design and development is reasonable, given other priorities and commitments. Having approved the new product concept, and the associated risks that need to be taken, management should promptly allocate funds and appropriate human resources to support the next phase of design and development. Senior management should publicize throughout the company the priority of this project to signal the level of attention that it should command from all participants.

Senior management can then further facilitate the execution of the NPI effort by ensuring that the project team feels empowered to do its work of design and development. To do this, they must resist calling all the plays and eliminating choices for the team. While management needs to review the project at subsequent, prespecified checkpoints, they should refrain from introducing new design options or imposing their personal preferences during these later phases of the project.

They can accomplish this by:

Promoting the building of long-term infrastructure, including the capabilities of people and technology, and processes for the design, development, and manufacture of new products.

Establishing overarching product and technology strategies to guide the design of any new product from the beginning. Senior management will then be less likely to create strategy through forcing major design changes midway through the NPI project.

Reviewing early (before and at Gate 0) the apparent feasibility of the new product idea, and establishing or approving project targets (which may subsequently be refined or modified in Phase 1).

Holding projects to these targets, or consciously changing the goalposts when it is apparent that they cannot all be reached.

Clearly communicating how the current project fits existing company strategies.

By engaging in these activities, senior management can increase its positive influence over projects as opposed to today's common practice of hindering the process by interfering late in the game. By letting go of decision making (though not of ultimate control) during project implementation, senior management empowers the team. The fascinating and important paradox here is that senior management can increase their own influence over the project and simultaneously shift power to the team by accelerating the timing of their attention and elevating the content of the issues with which they are concerned. Thus, empowered teams become the vehicle for cutting lead times and increasing customer satisfaction.

FACILITATE HORIZONTAL INTEGRATION THROUGH CODEVELOPMENT

Companies get much faster at NPI without compromising other objectives, and without falling into big traps, when they learn how to facilitate integrated horizontal work; that is, work among the experts who need to take part in various aspects of product design and development. This can be accomplished by setting targets for the project as a whole, building collaborative teams that pursue product objectives, and improving the team's productivity in conducting the project.

The core team identifies common targets for the project by establishing objectives for product performance, other aspects of quality, unit cost, and development time and cost. They must balance this set of targets and specify from the beginning the priorities among them. The managerial challenges here are:

Employing all existing knowledge of requirements and capabilities.

Establishing these targets simultaneously based on external and internal requirements and capabilities.

Communicating a consistent message about these targets and their interrelationships, and assuring that all team members embrace the product objectives.

Managing the NPI process to achieve the target-set, and in the face of slippage of one type or another knowing how to best revise all of the targets.

Assessing (in retrospect) strengths and weaknesses in pursuing the targets, as a basis for achieving the targets of future projects.

To achieve collaborative behavior from cross-functional teams, real communication needs to develop among specialists grounded in different subcultures. Effective collaboration requires surfacing the cultural differences that are getting in the way, and thus reducing barriers, while also building a new shared culture among team members. For collaboration to be the result, the new culture needs to be one in which it is even necessary to share problems and failures, and to take risks, because members understand that the emphasis will be on how to fix and learn rather than on pointing fingers or making threats. Repeated, direct, and early exposure of all potential NPI team members to the requirements of customers is one strategy for enhancing collaborative behavior. This can be accomplished by having all key team members spend time observing their current products in use and hearing about desired improvements, and by employing structured design methodologies such as quality function deployment. Another step toward true collaboration is to establish an in-house training program designed to strengthen team synergy and joint problem solving.

Effective collaboration is also determined, in part, by the initial selection of team members. This is a more complex matter than often realized. Companies tend to put various specialists together and hope they have created the needed synergy. But there is more to team selection than this. A more sophisticated approach is to look not only for technical skills and representation from all affected areas, but also for sufficient prior NPI experience, knowledge of company practices and procedures, and (for a major product introduction) a project manager with seniority. In addition to all of these considerations, companies must also assess potential team members and the project manager in terms of their interpersonal skills and problem-solving styles. Once a strong team is assembled, it is important for core team members to stay on the project until its completion.

Core teams rely on other organizational units, both within the company (such as test, quality, or volume manufacturing) and outside the company (such as suppliers who have design responsibilities). It is not enough, therefore, that there be synergy within the core team. Members of the core team must also be effective in their dealings with others, who ought to be considered as part of the extended team, even though their primary work orientation may lie elsewhere.

NPI teams can cut lead times and increase customer satisfaction by overlapping streams of related work; that is, starting downstream development work before work on the prior phase is completed. This transition from sequential to concurrent work typically begins with overlapping product and process development and bringing manufacturing engineering into the development process much earlier. Such overlapping—commonly referred to as concurrent or simultaneous engineering, or, simply, codevelopment—directly cuts lead times. It also facilitates the use of design for manufacture (DFM) techniques which can reduce total lead times even further by eliminating the need for major redesigns and facilitating rapid manufacturing ramp-up.

As the codevelopment process matures through changes to the work structure and problem-solving activity, several other benefits may accrue. First, the NPI projects expand to include more players, such as suppliers, and people from purchasing and field services. Suppliers are often on the critical path of an NPI project, so bringing them in early (along with those who will arrange for contracting with them) can directly shorten the required lead time. Field service personnel have firsthand knowledge of customer patterns of use, operating problems with existing products, and unmet needs; they can promote design choices that will satisfy these requirements (and those of serviceability of the product).

Through this transition to codevelopment, a communication process evolves from being reactive to interactive, and finally to the proactive inclusion of others. The quality and tone of the working relationship shifts from the neglect of others, to the solicitation of specifications from others ("What should I have from you to do my piece better?"), to one of true partnership: "How can we work together to do the best possible job?"

This switch from a "me" attitude to a "we" attitude supports the reduction of lead time by encouraging the right thing to be done the first time, and thereby eliminating time-consuming and costly rework. Even when there is little or no time pressure on a project, codevelopment will promote effective product design and development with appropriate kinds of risk-taking.

When a company begins to implement such changes, it is likely to realize that it is on a path of continuous improvement. As a company moves forward from its traditional sequential and functionally oriented approaches, most managers will notice progress. But as managers reflect on their evolving mind-sets and relationships, they will see even greater possibilities for taking their change processes further. What they may have considered to be codevelopment will then appear only as an early and incomplete transitional state.

When all the members of an NPI team are working on the same integrated design, rather than different pieces of design, a harmonious product is more likely to result. This can only be achieved when natural barriers to communication—different thought-worlds, assumptions and meanings—have been overcome. The resulting product will then do more than perform well and provide outstanding value; it will also be perceived by the customer as having coherence and being distinct.

USE CUSTOMER REQUIREMENTS TO DRIVE PROJECT MANAGEMENT

Throughout the project, the full team must strive to include the customer's point of view in all decisions of design and development. From the beginning, customer needs should be focused, limited in number, prioritized, and clear. The needs of customers should be reconciled with technological possibilities and capabilities to establish clear technology requirements early in the project's life. These technology requirements are the foundation for downstream vision and focus. While there is always the danger of creating products that will not be distinct from others (me-too products), there is an even greater danger of being unrealistically ambitious in setting technology requirements. In order not to sink or unnecessarily slow down a project, managers should not overstate requirements for either product or associated manufacturing technology; a clear understanding of what is essential to customers can help to calibrate those requirements.

Choosing and staying focused on the right design problems becomes essential if shortened lead times and increased customer satisfaction is to result. Competence in the use of structured design tools, model building, methods of rapid prototyping, and other aids to visualization further a common understanding of the intended product. Planning for reusability and commonality in product designs will shorten lead times compared with the frequent tendency of designers to prefer to generate "clean sheet" product designs. Product designers also must resist the tendency of creeping elegance, where unnecessary features or performance attributes are pursued and become the cause of late introductions. The effective integration of industrial design capabilities can provide a special advantage, especially when competitors are evenly matched in technological capabilities.

Lead times can also be cut through actions that reduce the inherent variability in an NPI project. Aside from the development of new technologies during the project, other common sources of time variability are: tooling development; vendor start-up; software development; testing; regulatory compliance; new organizational structures; and human resource recruitment or development. While it is difficult to predict exactly which bottlenecks will materialize, it is possible to plan so that their materialization does not disrupt the entire program. Some fruitful approaches— which, admittedly, are not always feasible—are:

Contingency plans to back up critical path items having high perceived variance.

Off-line development of advanced product and process technologies to produce relatively mature capabilities that can readily be applied to new projects when needed.

Off-line development of basic infrastructure, such as technologies that enhance product design (CAD and EDI, for example) and training in their use.

Project managers should play a key role in the execution of a new product development effort. Their contributions should include:

Promoting the shared product vision that supports horizontal integration.

Resisting optimistic assumptions that generate unpleasant surprises and invite subsequent defensive behavior.

Watching for opportunities to improve the productivity of individuals and the project as a whole.

Providing a critical interface with management and "buffer" with the rest of the organization.

In particular, the project manager can help ensure that the project is benefiting from external enablers elsewhere in the company (e.g., TQM, human resource development, and information technology) and that critical linkages with other operating groups (e.g., marketing, field service, and manufacturing process development) are functioning as they should. In short, the project manager should facilitate team members in doing their best work toward a common end, and also help them learn how to get even better.

CONCLUSION

Working together, senior management, project management and those who specialize in various design and development activities, can reshape the NPI process to be more effective. Although numerous companies have reported breakthrough product introductions by using a small, one-shot, specially selected, colocated team (called *skunkworks* or *tiger teams*) this approach is not a long-term answer for most U.S. companies. To achieve a rhythm of successful new product introductions, a solution that modifies, rather than avoids, the existing organizational structure is required.

There is usually no single quick fix to problems of excessive lead times or products that are not competitive. When companies seek to restructure their product development process, they must be careful to recognize the limitations of adopting processes that were used successfully elsewhere. Blind benchmarking can lead to counterproductive initiatives. Other successful companies may not have comparable contexts for product design and development (type of market, position in industry, strategies, culture, and competencies). Furthermore, formal processes rarely tell the whole story of a company's success in bringing new products to market. Many problems in cutting lead time and increasing customer satisfaction can be traced to the execution, rather than the formulation, of policies and procedures. Excessive reliance on management meetings and administrative reports, unproductive team meetings, and wasted individual work are mundane but deadly barriers to swift project execution.

Achieving discipline in the execution of product design and development within a simplified ongoing structure should be a dominant objective. Besides the delays, confusions, and need for rework already mentioned, the modification of product specifications during an NPI project is one of the most common sources of excessive delay in product introduction. Companies with unacceptably long, new product lead times and extensive changes in product definition throughout the project should ask themselves: Do we keep our product specifications flexible because we are in a long lead time business, or is our development time unacceptably long because we allow our product specifications to be flexible?

While faster development cycles are important for companies to achieve, not all product development projects are under the same degree of time pressure. There will be a point beyond which speed becomes reckless. Management should resist the tendency to mandate an unnecessarily short lead time that could result in products being poorly designed and shipped before problems are discovered and remedied.

In summary, cutting lead time and increasing customer satisfaction requires strengthening a company's capability for horizontal integration, so that a cross-functional team can work effectively and productively toward a common end. Management can accelerate such improvements by promoting joint responsibility for joint action, and nurturing a climate where continuous improvement is pursued by all. Ultimately, however, success will come from relentless discipline in listening for the voice of tomorrow's customer and shunning the unproductive practices of the past.

The assessment of completed projects offers valuable opportunities for organizational learning, some elements of which will be: how to choose and sequence the many possible improvement initiatives, use new performance measures to provide incentives for desired behavior, select and manage teams, and integrate suppliers and customers in the NPI process. Other elements of learning will be more closely tied to the strategy of product development: how to define product families, routinely enhance existing products, better anticipate the needs of customers, and introduce new technology. This journey will be similar for many companies, but the path will vary. Most important for any single company is to blaze its own path with vigor, adventure, accomplishment, and reflection.

BIBLIOGRAPHY

Adler, Paul S.; Henry E. Riggs; and Steven C. Wheelwright. "Product Development Know-How: Trading Tactics for Strategy." *Sloan Management Review* 31, no. 1, Fall 1989.

Akao, Y., and M. Kogure. "Quality Function Deployment and CWQC in Japan." *Quality Progress* 16, no. 10, October 1983, pp. 25–29.

Andreasen, M. Myrup; S. Kahler; and T. Lund. *Design for Assembly*. New York: Springer-Verlag, 1983.

Argyris, Chris. "Teaching Smart People How to Learn." *Harvard Business Review*, May–June 1991, pp. 99–109.

Badawy, M. K. "Understanding the Role Orientation of Scientists and Engineers." *Personnel Journal*, June 1971, pp. 449–85.

Bebb, H. Barry. "Implementation of Concurrent Engineering Practices." *Proceedings of the Society of Manufacturing Engineers, Concurrent Engineering Design Conference, Dearborn, Michigan, 1990.*

Bennett, Amanda. "Company School: As Pool of Skilled Help Tightens, Firms Move to Broaden Their Role." *The Wall Street Journal*, May 8, 1989.

Berg, Jeffrey, and Jeffrey Loeb. "The Role of Field Service in New Product Development and Introduction." *AFSMI Journal*, May 1990.

Blackburn, Joseph D. *Time-Based Competition*. Homewood, Ill.: Business One Irwin, 1991.

Boike, Douglas, and Stephen R. Rosenthal. "Managing Technology in New Product Introduction." *Boston University Manufacturing Roundtable*, August 1990.

Boothroyd, G., and P. Dewhurst. *Product Design for Assembly*. Wakefield, R.I.: Boothroyd Dewhurst Incorporated, 1987.

Clark, Kim B., and Takahiro Fujimoto. *Product Development Performance: Strategy, Organization and Management in the World Auto Industry*. Boston: Harvard Business School Press, 1991.

Clausing, Don, and Stuart Pugh. "Enhanced QFD." *Proceedings of the Design Productivity Institute International Conference*, Honolulu, February 1991.

Cooper, Robert G. "Stage-Gate Systems: A New Tool for Managing New Products." *Business Horizons* 33, no. 3, May–June 1990, pp. 44–54.

Crosby, Philip B. *Quality Is Free*, New York: McGraw-Hill, 1979.

Davis, Stanley M. *Managing Corporate Culture*. Cambridge, Mass.: Ballinger, 1984.

Dehner, Jerry. "The Design and Development of Northern Telecom's Meridian Norstar Business Phone System." *Boston University Manufacturing Roundtable*, July 1990.

Dertouzos, M.; R. Lester; and R. Solow. *Made in America: Regaining the Competitive Edge*. Cambridge, Mass.: M.I.T. Press, 1989.

Dixon, J. Robb; Alfred J. Nanni; and Thomas E. Vollmann. *The New Performance Challenge*. Homewood, Ill.: Business One Irwin, 1990.

Dougherty, Deborah. *Interpretive Barriers to Successful Product Innovation*, Report No. 89–114. Cambridge, Mass.: Marketing Science Institute, September 1989.

Ealey, Lance, and Leif G. Soderberg. "How Honda Cures 'Design Amnesia.'" *The McKinley Quarterly*. Spring 1990, pp. 3–14.

Ebner, Merrill, L. "The New Product Introduction Process: The Role of the Manufacturing Manager." *Boston University Manufacturing Roundtable*, August 1990.

Ettlie, John E., and Henry W. Stoll, eds. *Managing Design in Manufacturing*. New York: McGraw-Hill, 1990.

Foster, Richard. *Innovation: The Attacker's Advantage*. New York: Summit Books, 1986.

Francis, Dave, and Don Young. *Improving Work Groups: A Practical Manual for Team Building*. San Diego: University Associates, Inc., 1979.

Garvin, David. "Competing on the Eight Dimensions of Quality." *Harvard Business Review*, December 1987.

Gentile, A. Liza, and Stephen R. Rosenthal. "The Design and Development of NeXT's Computer System." *Boston University Manufacturing Roundtable*, December 1989.

Gerwin, Donald. "Do's and Dont's of Computerized Manufacturing." *Harvard Business Review*, March–April 1982, pp. 107–16.

Ginn, Martin E. "Key Organizational and Performance Factors Relating to the R&D/ Production Interface." PH.D. dissertation, Northwestern University, Department of I.E.M.S. Evanston, Ill., 1983.

Ginn, Martin E., and Albert H. Rubenstein. "The R&D/Production Interface: A Case of New Product Commercialization." *Journal of Product Innovation Management* 3, 1986, pp. 158–70.

Graham, Margaret B. W., and Stephen R. Rosenthal. "Flexible Manufacturing Systems Require Flexible People." *Human Systems Management* 6, 1986, pp. 211–22.

Gregory, Donald. "The Design and Development of GE's CF6-80A Aircraft Engine." *Boston University Manufacturing Roundtable*, August 1990.

Gregory, Kathleen L. "Native-View Paradigms: Multiple Cultures and Culture Conflicts in Organizations." *Administrative Science Quarterly* 29, 1983, pp. 359–76.

Gupta, Ashok K.; S. P. Raj; and David Wilemon. "The R&D-Marketing Interface in High-Technology Firms." *Journal of Product Innovation Management* 2, 1985, pp. 12–24.

Hauser, John R., and Don Clausing. "The House of Quality." *Harvard Business Review*, May–June 1988, pp. 63–73.

Hayes, Robert H.; Steven C. Wheelwright; and Kim B. Clark. *Dynamic Manufacturing*. New York: The Free Press, 1988, chaps. 10 and 11.

Heskett, John. "Integrating Design into Industry." *Design Processes Newsletter* 3, no. 2, 1989.

Jaikumar, R. "Post-Industrial Manufacturing." *Harvard Business Review*, November–December 1986a, pp. 69–76.

Jaikumar, R. "Hitachi Seiki (A)." *Harvard Business School Case* 9-686-103 1986b.

Juran, J. M. *Juran on Leadership for Quality*. New York: The Free Press, 1989.

Katz, Robert, and Michael Tushman. "Communication Patterns, Project Performance, and Task Characteristics: An Empirical Evaluation and Integration in an R&D Setting." *Organization Behavior and Human Performance* 23, 1979, pp. 139–62.

Kerzner, Harold. "Project Management: A Systems Approach to Planning, Scheduling, and Controlling." New York: Van Nostrand Reinhold, 1989.

King, Bob. *Better Designs in Half the Time: Implementing QFD in America*. 3rd ed. Methuen, Mass.: Goal/QPC, 1989.

Kirkland, Carl. "Meet Two Architects of Design-Integrated Manufacturing." *Plastics World*, December 1988.

Kraus, William A. Collaboration in Organizations: Alternatives to Hierarchy. New York: Human Sciences Press, 1980.

Liker, J. K., and W. M. Hancock. "Organizational Systems Barriers to Engineering Effectiveness." *IEEE Transactions on Engineering Management* EM-33, no. 2, May 1986.

Lloyd, Frank. "The Design and Development of Motorola's Keynote Radio Paging Receiver." *Boston University Manufacturing Roundtable*, August 1990.

Maidique, M. A., and B. J. Zirger. "New Products Learning Cycle." *Research Policy*, December 1985, pp. 1–40.

Miller, Jeffrey G., and Jay S. Kim. "Beyond the Quality Revolution: U.S. Manufacturing Strategy in the 1990s." *Boston University Manufacturing Roundtable*, 1990.

Nevins, James L., and Daniel E. Whitney, eds. *Concurrent Design of Products and Processes*. New York: McGraw-Hill, 1989.

Nussbaum, Bruce. "Needed: Human Capital." *Business Week*, September 19, 1988, p. 100–3.

Ohmae, Kenichi. *The Borderless World*. New York: Harper Collins, 1990.

Pugh, Stuart. "Concept Selection—A Method that Works." *International Conference on Engineering Design (ICED 81)*, Rome, Italy, 1981.

Raelin, Joseph A. *The Clash of Cultures: Managers and Professionals*. Boston: Harvard Business School Press, 1985.

Rosenthal, Stephen R. "Bridging the Cultures of Engineers; Challenges in Organizing for Manufacturable Product Design." Chapter 2 in *Managing Design in Manufacturing*. Edited by John E. Ettlie and Henry Stoll. New York: McGraw-Hill, 1990a. (Also available as a working paper of *Boston University Manufacturing Roundtable*, August 1989.)

Rosenthal, Stephen R. "Building a Workplace Culture to Support New Product Introduction." *Boston University Manufacturing Roundtable*, September 1990b.

Rosenthal, Stephen R., and Harold Salzman. "Hard Choices about Software: The Pitfalls of Procurement." *Sloan Management Review*, Summer 1990, pp. 81–91.

Rosenthal, Stephen R., and Mohan V. Tatikonda. "Managing the Time Dimension in the New Product Development Cycle." *Boston University Manufacturing Roundtable*, August 1990.

Rosenthal, Stephen R., and Mohan V. Tatikonda. "Competitive Advantage through Design Tools and Practices." In *Integrating Design and Manufacturing for Competitive Advantage*. Edited by Gerald I. Susman. New York: Oxford University Press, 1992.

Shunk, Daniel. *Integrated Process Design and Development*. Homewood, Ill.: Business One Irwin, 1992.

Skinner, Wickham. "The Focused Factory." *Harvard Business Review*, May–June 1974, pp. 113–21.

Skinner, Wickham. "The Productivity Paradox." *Harvard Business Review*, July–August 1986, pp. 55–59.

Smith, Larry. "QFD Implementation at Ford." *Concurrent Engineering Design Conference Proceedings*. Dearborn, Mich.: Society of Manufacturing Engineers, 1990.

Souder, William E. *Managing New Product Innovations*. Lexington, Mass.: D. C. Heath, 1987

Stalk, George, and Thomas M. Hout. *Competing against Time*. New York: The Free Press, 1990.

Stoll, Henry W. "Design for Manufacture." *Manufacturing Engineering*, January 1988, pp. 67–73.

Tatikonda, Mohan V., and Stephen R. Rosenthal. "The Design and Development of AGFA Compugraphic's CG9400 Imagesetter." *Boston University Manufacturing Roundtable*, 1990.

Taylor, James C.; Paul W. Gustavson; and William S. Carter. "Integrating the Social and Technical Systems in Organizations." *Managing Technological Innovation*. Edited by Donald D. Davis. San Francisco: Jossey-Bass, 1986, pp. 154–86.

Thamhein, Hans J. "Managing Technologically Innovative Team Efforts toward New Product Success." *Journal of Product Innovation Management* 7, no. 15, 1990, pp. 5–18.

Tushman, Michael. "Technical Communication in R&D Laboratories: The Impact of Project Work Characteristics." *Academy of Management Journal* 21, no. 4, 1978, pp. 624–45.

Urban, Glen L.; John R. Hauser; and Nikhilesh Dholakia. *Essentials of New Product Management*. Englewood Cliffs, N.J.: Prentice Hall, 1987.

Urban, Glen L., and Eric Von Hippel. "Lead User Analyses for the Development of New Industrial Products." *Management Science* 34, no. 5, May 1988, pp. 569–82.

Venugopal, K. C. "The Design and Development of Amdahl Corporation's 5890 Mainframe Computer." *Boston University Manufacturing Roundtable*, 1990.

Wasserman, Arnold S. "Redesigning Xerox: A Design Strategy Based on Operability." In *Ergonomics*, edited by Arnold S. Wasserman. Norwood, N.J.: Ablex Publishing Corp., 1989, pp. 7–44.

Welter, Therese R. "Designing for Manufacture and Assembly." *Industry Week*, September 4, 1989.

Westney, D. Eleanor, and Kiyonori Sakakibara. "Designing the Designers." *Technology Review*, April 1986.

Wheelwright, Steven C., and W. E. Sasser, Jr. "The New Product Development Map." *Harvard Business Review*, May–June 1989.

Whitton, Tony, and Becky Cook. "The Design and Development of Interlake Conveyors' 'Knock Down' Selecta-Flow Shelf." *Boston University Manufacturing Roundtable*, 1990.

Wolff, Michael. "Building Teams—What Works (Sometimes)." *Harvard Business Review*, November–December 1989, pp. 9–10.

GLOSSARY

AME Advanced manufacturing engineering

BOM Bills of material

BOP Business opportunity plan

CAD Computer-aided design

CAE Computer-aided engineering

CAM Computer-aided manufacturing

CAPP Computer-aided process planning

CASE Computer-aided software engineering

CID Computer-integrated design

CIM Computer-integrated manufacturing

DFA Design for assembly

DFM Design for manufacture

DOC Direct operating cost

DPU Defect per unit

ECO Engineering change order

EDI Electronic data interchange

EEPROM Electrically erasable programmable read only memory

E-mail Electronic-mail network

EMI Electromagnetic interference

ETOP Extended twin operations

FETT First engine to test

FMS Flexible manufacturing systems

FOM Figures of merit

GE General Electric Co.

GEAE GE Aircraft Engine

GT Group Technology

HPT High-pressure turbine

HR Human resource

IPO Initial production organization

IT Information technology

JIT Just-in-time

KSU Key service unit

LCC Life cycle cost

LCD Liquid crystal display

LPT Low-pressure turbine

ME Manufacturing engineering; *also* Mechanical engineer

MTBF Mean time between failure

N/C Numerical control

NPI New product introduction

NTL Northern Telecom Ltd.

OEM Original equipment manufacturer

PBX Private branch exchange

PCB Printed circuit board

PERT Project evaluation review technique

PPM Parts per million

QC Quality control

QFD Quality function deployment

RFI Radio frequency interference

SFC Specific fuel consumption

Six Sigma   A measure of goodness—the capability of a process to produce perfect work

SMT   Surface-mount technology

SPC   Statistical process control

TMF   Turbine midframe

TQM   Total quality management

VLSI   Very large-scale integration

VPE   Value/process engineer

WBS   Work breakdown structure

INDEX

OTHER TITLES IN THE BUSINESS ONE IRWIN/APICS LIBRARY OF INTEGRATIVE RESOURCE MANAGEMENT

Integrated Process Design and Development
Dan L. Shunk

Shows how to design and develop processes that are consistent with the capabilities of the plant and the employees. Shunk introduces process design terminology in reader-friendly terms instead of using the complex jargon found in many manufacturing books.

ISBN: 1-55623-556-9 $42.50

Marketing for the Manufacturer
J. Paul Peter

Explains the marketing role to the nonmarketing specialist. Peter provides a detailed analysis of how marketing fits into various organizational structures and product management systems. He offers methods for researching consumer markets and creating a dynamic strategic plan.

ISBN: 1-55623-648-4 $42.50

Integrated Production and Inventory Management
Thomas E. Vollmann, D. Clay Whybark, and William L. Berry

Explains the use of modern planning and control systems to remove excess inventory investment, save on production and distribution costs, and better integrate organization efforts. The authors give you strategies and systems to optimize customer service through the use of the latest inventory monitoring procedures.

ISBN: 1-55623-604-2 $42.50

Managing for Quality
Integrating Quality and Business Strategy
Howard S. Gitlow

Details how to integrate quality into the heart of a company's business plan and use it to gain a strategic edge over the competition. Gitlow shows how to satisfy customer requirements and reduce the cost of quality. He offers methods for detecting and preventing defects by using a system that minimizes monitoring efforts.

ISBN: 1-55623-544-5 $42.50

Managing Human Resources
Integrating People and Business Strategy
Lloyd S. Baird

Details specific steps for integrating the human resource function into all other key areas, including customer relations and field service. Baird shows how the entire organization is affected by the actions of each of its workers. He offers strategies for achieving human resource goals at the line management level.

ISBN: 1-55623-543-7 $42.50

Prices quoted are in U.S. currency and are subject to change without notice.

Available in fine bookstores and libraries everywhere.